UNIVERSITY–INDUSTRY RESEARCH PARTNERSHIPS

Recent Titles from QUORUM BOOKS

Revolution in the World Petroleum Market
Mary Ann Tétreault

Competition and Diversification in the United States Petroleum Industry
Arabinda Ghosh

Productivity and Quality Through People: Practices of Well-Managed Companies
Y. K. Shetty and Vernon M. Buehler, editors

Public Policy and the Practice and Problems of Accounting
Ahmed Belkaoui

Boards of Directors: Their Changing Roles, Structure, and Information Needs
Charles N. Waldo

A Compendium of State Statutes and International Treaties in Trust and Estate Law
M. Henner

Utilizing Consultants Successfully: A Guide for Management in Business, Government, the Arts and Professions
Herman R. Holtz

International Broadcasting by Satellite: Issues of Regulating Barriers to Communication
Jon T. Powell

Corporate and Commercial Free Speech: First Amendment Protection of Expression in Business
Edwin P. Rome and William H. Roberts

The Impact of Office Automation on Clerical Employment, 1985–2000: Forecasting Techniques and Plausible Futures in Banking and Insurance
J. David Roessner, Robert M. Mason, Alan L. Porter, Frederick A. Rossini, A. Perry Schwartz, and Keith R. Nelms

The Learning Curve: A Management Accounting Tool
Ahmed Belkaoui

Willing Workers: The Work Ethics in Japan, England, and the United States
Tamotsu Sengoku, translated by Koichi Ezaki and Yuko Ezaki

University–Industry Research Partnerships

THE MAJOR LEGAL
ISSUES IN RESEARCH
AND DEVELOPMENT AGREEMENTS

Bernard D. Reams, Jr.

Q

QUORUM BOOKS

WESTPORT, CONNECTICUT • LONDON, ENGLAND

Library of Congress Cataloging-in-Publication Data

Reams, Bernard D.
University-industry research partnerships.

Bibliography: p.
Includes index.
1. Research and development partnership—United States. 2. Research and development contracts—United States. I. Title.
KF1380.7.R43 1986 346.73'023 85-9589
347.30623
ISBN 0-89930-121-5 (lib. bdg. : alk. paper)

Library of Congress Catalog Card Number: 85-9589
ISBN: 0-89930-121-5

First published in 1986 by Quorum Books

Greenwood Press
A division of Congressional Information Service, Inc.
88 Post Road West, Westport, Connecticut 06881

Printed in the United States of America

The paper used in this book complies with the Permanent Paper Standard issued by the National Information Standards Organization (Z39.48-1984).

10 9 8 7 6 5 4 3 2 1

Copyright Acknowledgments

Permission from authors and publishers to quote from the following texts is gratefully acknowledged:

Derek Bok, *Beyond the Ivory Tower: Social Responsibilities of the Modern University* (Cambridge: Harvard University Press, 1982). All quotations are reprinted by permission of Harvard University Press. Copyright © 1982 by the President and Fellows of Harvard College.

Donald R. Fowler, "A Study of the Need for and the Impediments to Improved and Novel University-Industry Research Relationships" (Ph.D. dissertation, Claremont Graduate School, copyright © 1983 by Donald R. Fowler). All quotations are reprinted by permission of the copyright holder.

Donald R. Fowler, "University-Industry Research Relationships: The Research Agreement," 515-532. *Journal of College and University Law*, 9 (1982-1983). Permission given by the National Association of College and University Attorneys as the copyright holder.

Thomas W. Langfitt, Sheldon Hackney, Alfred P. Fishman, and Albert V. Glowasky, eds., *Partners in the Research Enterprise: University-Corporate Relations in Science and Technology* (Philadelphia: University of Pennsylvania Press, 1983). All quotations are reprinted by permission of the University of Pennsylvania Press and are copyrighted 1983 by the University of Pennsylvania Press.

To
Bridget, Andrew, and Adriane

The sciences having long seen their votaries labouring for the benefit of mankind without reward, put up their petition to Jupiter for a more equitable distribution of riches and honours. . . . A synod of the celestials was therefore convened, in which it was resolved, that patronage should descend to the assistance of the sciences. Patronage was the daughter of Astrea, by a mortal father, and had been educated in the school of truth, by the goddesses, whom she was now appointed to protect. . . . She came down, with the general acclamation of all the powers that favour learning. Hope danced before her, and liberality stood at her side, ready to scatter, by her direction, the gifts which fortune, who followed her, was commanded to supply.

—Samuel Johnson, *The Works of Samuel Johnson, LL.D.*
Vol. 2: *The Rambler No. 91, Tuesday, January 29, 1751*
(London: Oxford University Press, 1825), p. 426.

Contents

Preface

Within the past decade rapid scientific breakthroughs in university sponsored biotechnology research have attracted the interest of corporations which see these scientific advancements resulting in the promise of big profits. Historically, universities have maintained close ties with industry, whether it be for philanthropic support, use of faculty as consultants, or the funding of research partnerships. However, over the past few years, with the advent of large, multimillion dollar biomedical contracts between industry and universities, concern has been expressed regarding the ethical and legal issues which confront the nonprofit university entering into these new agreements. As an introduction, this book identifies and pulls together a variety of historical sources precedent to today's major university–industry contractual arrangements. Beginning with the Morrill Act to the present, the research policies of government, industry, and universities are reviewed.

This text examines the legal problems inherent in university–industry cooperation and the method in which our laws are utilized by both parties to address those concerns. The law of university–industry research relations is categorized and examined under the following topics: contract law and the case law of higher education, general contract law and business law, contract law and statutory guidelines, and methods of dispute resolution. Major judicial cases, statutes, and administrative rules and regulations are cited and discussed.

An original analytical and comparative examination of three major research contracts is presented: Massachusetts General Hospital–Hoechst A.G. Contract of 1980 ($70 million); Washington University–Mallinckrodt, Inc., Agreement of 1981 ($3.88 million); Washington University–Monsanto Biomedical Research Agreement of 1982 ($23.5 million). A fourth contract—Washington University–Anheuser-Busch Micromixing Research Agreement of 1983 ($600,000)—concludes the presentation as an example of a typical spin-off agreement resulting from prior successful relationships with industry.

The success of these existing agreements serves as the foundation for future cooperative research ventures. Many other major research corporations are watching the results of these research interactions before seeking university ties. Foreign corporations appear to be eager to get positioned at a window on American academic research. In January 1985, the Italian drug company FIDIA Pharmaceuticals agreed to give Georgetown University at least $62 million over twenty years to establish a research institute to study the working of the brain. In examining future research options it is evident that the most promising appears to be the research and development limited partnership (RDLP). The RDLP allows research projects to be conceived and conducted in the normal university manner independent of outside intervention. The university gets to share substantially in the profits of successful products, and the tax laws make the RDLP an attractive investment vehicle for venture capitalists and investors.

Other options being adopted by institutions of higher education include creation of the university's own technology transfer organizations in cooperation with private venture capitalists. The advantages and disadvantages of these legal concepts are presented. The success of university–industry research ventures holds promise for improving our national research efforts and for society's betterment. Certainly the role of higher education in both basic and applied research may be profoundly affected.

Acknowledgments

A research project such as this could not have reached closure without the advice of a variety of persons experienced in cooperative research ventures. The suggestions and personal statements provided by Edward L. MacCordy, Associate Vice Chancellor for Research at Washington University, and his assistant Harry S. Leahey were invaluable in providing insight into the spirit and the reality of the research agreements discussed in this book. Tom L. Tolbert of Monsanto was most helpful in providing personal observations on the corporate viewpoint. Peter H. Ruger, General Counsel for Washington University, contributed time and energy in reading various sections for their accuracy on points of law and in suggesting areas for expansion. The author recognizes the patience of his family, from whom many hours were taken so that this project could reach its conclusion. To them this book is affectionately dedicated.

Appreciation is expressed to those individuals who encouraged and offered advice on this research project. A special thanks goes to Charles Michael Stanton, Jerome J. Marchetti, S.J., and Gerard A. Fowler, all of Saint Louis University, who read revision after revision. No research project is complete without the able assistance of librarians. Margaret H. McDermott and Margaret A. Goldblatt, both professional law librarians at Washington University's Freund Law Library, lent valuable assistance, as did Administrative Assistant Lucy C. Hubert.

The Association for the Study of Higher Education (ASHE) selected the original text of this book as first runner-up in its Outstanding Dissertations in Higher Education—1983–1984 Awards. The author appreciates this recognition by ASHE.

UNIVERSITY–INDUSTRY RESEARCH PARTNERSHIPS

1

The Problems of Collaborative Research

The primary aim of universities is to spread higher learning and to provide both the foundations and the technical knowledge for the learned professions. Since propagation of true learning is impossible without original investigation, an equally important function of American universities has been to conduct scientific research.

The question of whether the primary function of the university should be teaching or research is an historic one.[1] In referencing the expectation that scientific investigation would occupy a significant aspect of the newly created University of Virginia, Thomas Jefferson wrote: "With respect to the degree in which sciences will be taught here, I think I may say in as high an [*sic*] one as in the universities of Europe, should any of the students propose to pursue them so far."[2]

The introduction of academic specializations and the addition of professional education into the university structure have allowed universities to place special emphasis on technical specializations and research. The university is more than a research institution and its function is greater than that of affording adequate technical equipment to the professionals it trains; its basic purpose is to raise the intellectual and cultural standards of society.[3]

Historians Richard Hofstadter and C. Dewitt Hardy recognized the contemporary role the university plays in improving the human condition through useful research.

> One of the primary distinctions of the modern university is its ability to contribute actively to the sum of human knowledge. Research . . . is one of the pleasurable functions of the well-trained mind in any field, and it can be justified on the loftiest intellectual and humanistic grounds, as well as the more obvious ground that university research has become an indispensable part of the practical work of modern life.[4]

In the United States scientific research of all kinds has traditionally been carried on largely in and through universities and operating government bureaus, rather than through independent institutes or academies of science. The largest amount of scientific research has been conducted in the universities, although an increasing share of the cost of research has been borne by the government.[5]

> Modern scholarship in general and scientific method in particular have transformed research into something like a profession. . . . In a way it is a singular thing that research should have become a major activity of higher education. It became a part of it only within the last hundred years. Since then research has moved steadily to the foreground. In the past professors with curiosity made contributions to man's knowledge; only recently has productive research come to be considered essential to their position. . . . The major factor in bringing this situation about has been the development of the scientific method of discovery.[6]

Among performers of research the most important institution is the college or university. "In the United States it is estimated that 80% of all basic research is accomplished in universities; most of the remainder is in large-equipment national laboratories."[7]

From an educational perspective, research by institutions of higher learning provides an investment with unparalleled fringe benefits. Not only do research projects uncover new scientific knowledge and solve practical problems, but they also provide training grounds for new generations of scientists and engineers. Further, instructional laboratories in expensive high-technology areas are largely dependent on equipment inherited from sponsored research projects to supplement their budgetary allocations.[8]

In both private and public universities, funds for grants or contracts for research may come from the federal government, from industrial corporations, from private foundations, labor unions, other non-profit organizations, or from private individuals. Universities and federal government research units account for nearly four-fifths of the funds spent for social science research in the United States. Of the remaining one-fifth, only a small amount is spent by private foundations, which play a very minor role as performers.

Richard G. Cunningham, Vice President for Research and Graduate Studies at Pennsylvania State University, has noted that during the past twenty-five years industry has played a relatively minor role in academic research. He observes that only five to ten per cent of university research expenditures come from industry sources.[9] With the current decline in government funding of academic research, institutions of higher education now see the need to shift to more industrial support to compensate for this funding loss.

Industry sponsored research is not the ultimate financial cure for university research funding needs, as business is most often interested in a relatively narrow range of subjects. Prime areas of joint undertakings exist with genetic engineering (or bio-chemical/bio-medical engineering or gene splicing) and with electronic engineering with integrated "chip" electronics. These technologies have a common problem in that they both require large and expensive equipment, are capital intensive, and, being on the forefront of technology, their equipment rapidly becomes obsolete. These cost factors are a major reason why universities have tended to follow rather than lead in the applied research areas.

The problem of scientific equipment deterioration in university research facilities was identified, and possible solutions suggested, by the Association of American Universities (A.A.U.) as a major issue confronting higher education. Recognizing a decade of obsolescence at university research facilities, the A.A.U. recommended direct yet balanced support for university facilities from the federal government, the states, and private industry. The following specific steps are suggested as starting points:

1. Key Executive agencies should develop a mechanism to periodically assess the condition of university research laboratories and the consequences for the nation of that condition. Such assessments should involve the Office of Science and Technology Policy, the National Science Foundation, the Department of Defense, the Department of Energy, the Department of Health and Human Services, the Department of Agriculture and the National Aeronautics and Space Administration. The views of industry and the universities also should be solicited. . . .

2. The facilities and equipment program earlier proposed to be undertaken by the National Science Foundation but deferred should be restored as a priority initiative in FY [Fiscal Year] 1983.

3. The Department of Defense, Department of Energy, the National Aeronautics and Space Administration, the Department of Health and Human Services, and the Department of Agriculture should establish research instrumentation and facilities rehabilitation programs targeted on the fields of science and engineering of primary significance to their missions. . . .

4. Incentives should be provided to encourage industries and the states to join with universities and the federal government in renewing the nation's university research laboratories.

5. Interest costs incurred in the construction, modernization, renovation, and repair of facilities in which federally supported research programs are carried out should be allowable indirect costs under OMB Circular A–21.[10]

One possible impact of industry sponsoring research with universities is that high equipment costs are offset or reduced.

As the government has cut back on supporting academic research, private industry has increasingly become an important source of funding and, consequently, has assumed a more significant role in medical and scientific research. In the past it has been difficult to estimate the amount spent by industrial performers on research because of the problem of deciding which activities are research (as distinguished from routine data collection for the purposes of making a marketing decision).[11] Other problems involved determining what funds are utilized for applied, basic, and developmental research needs. A major advantage offered industry by relying on the university as a research source for industrial studies rests in the following:

a centralized concentration of scholar-researchers; access to research equipment, laboratories, and facilities; centralized access to research literature in university libraries and data-bases.[12]

Studies by the New York University Center for Science and Technology Policy have identified the general divisions of corporate research donations to institutions of higher learning. Their two-year study of research interactions between industry and higher education found that engineering, medicine, and agriculture programs received about 86 percent of regular support provided by industry to universities. The center surveyed 465 university research programs of the major research universities and noted that 67 percent of that support went to engineering and computer science studies, while 14 percent went to basic medical and biological studies. Agricultural research received 5 percent of the total support.[13]

David F. Noble and Nancy E. Pfund identify three points in this century where universities have made major shifts in the funding sources for their scientific research. They note that universities typically undergo major shifts responding to pressures in the economic and political climate. Early in the twentieth century "elitist liberal-arts colleges" developed research centers for the newly emerging electrical and chemical industries. By the second World War university research ties shifted to the federal government in response to the need for military and agency contract research.

> Now, the universities are shifting their allegiance back to the private sector—and to the dominant power in that sector, the petrochemical industry—under the goad of grave financial problems and in an effort to escape from governmental red tape and scrutiny. The universities' new role will be to provide research and training in new industrial areas—particularly semiconductors, automation and biotechnology—and bestow ideological sanction and scientific legitimacy upon Big Business's campaign against government "interference" in the economy. In the vanguard of this shift are large research institutions like Harvard, Stanford, M.I.T. and the University of Michigan; leading the industry side are the major corporations in the petrochemical industry, Monsanto, Exxon, Dow, Du Pont and others. According to the annual survey of "The Chronicle of Higher Education," universities have, in the last three years, received the largest increase in financial support from industry since 1920.The latest transformation, like those before it, is bound to have far-reaching consequences for the direction of research and technology, patterns of funding and appointments, the form and content of higher education and the future of academic freedom.[14]

Sponsored research by private industry is now a source to be recognized as a significant factor in the funding of the goals of institutions of higher learning.

Robert M. Hutchins observed the modern development of scientific research at the American university with the following comment:

In 1936 research was having a hard time in the American university. People could understand the idea of teaching . . . [but] the mystique of science had not yet covered the earth.

Vast sums, unheard of in my day, are sought and obtained in the name of education. An increasing fraction of this money comes from government and business. And most of it, of course, goes for what is called research. Today whatever is done in a university that cannot be called teaching is called research.

The universities in fact are now engaged in three activities that are not very closely related to one another: research, vocational certification, and social accommodation. Research requires no further explanation, although it should be said that one who wants the universities to be centers of independent thought may well be alarmed at the conscious or unconscious lapses from independence that large-scale support from government and business may induce. The universities have demonstrated their willingness to do almost anything for money. Government and business are not wholly disinterested in their approaches to the universities: they are not seeking the truth, but are hiring universities to promote the ends they have in view. If the truth serves these ends, it is merely a coincidence.[15]

Edward Schaffer perceives the problem as one dividing various academic disciplines and pitting them against one another in competition for research funds. He identifies three types of clusters related directly to research motivation as it focuses on two social systems of action: purposive-rational action (technocracy) or symbolic interaction (culture). His basic categories of research are:

1. *The strict sciences* which produce or transmit instrumental knowledge used to contribute to the growth of production or the extension of the power of technical control.

2. *The historical-cultural-interpretive disciplines* which produce or transmit knowledge that yields an understanding of the social-cultural world and which aims at improving our understanding of symbolic interaction.

3. *The critical social sciences* which seek to criticize systematically ideologies with the explicit aim of attaining the ideal of a fully open society.

Problem-solving in the modern university is in danger of becoming overly identified with the research-guiding interest of the first discipline cluster. This discipline cluster or basic research category is replacing traditional discipline divisions in the university in importance since technocratic problem-solving requires cutting across traditional discipline boundaries. The strict science cluster contains both social and natural sciences, including technical specialties within so-called "humanistic" disciplines, which produce or transmit knowledge that can be used to enhance the power of technical control. . . .

The instrumental demands of industry, government and the occupational structure have split departments of academic disciplines into individuals and groups who identify with, or are opposed to, the technocratic norm of higher education. The advocates of technocratic problem-solving are often viewed by university administrators and educational planners as having greater academic legitimacy than their ivory tower counterparts. Educational planners and administrators who are sensitive to the instrumental demands of the wider society see the split within academic departments as justification for reorganizing disciplines around problem-solving courses with a systems orientation to inquiry.[16]

Yet despite these negative viewpoints, modern universities can claim that much of contemporary technology is the product of their work, directly or indirectly. This technological advancement has been achieved with responsibility. Of paramount importance to the university is its task to make life worth living, to make technology beneficial to mankind, to assist in directing society toward gaining a better civilization. Technology is one important measure of the advancement of a people.

Terry Sanford, President of Duke University, in his Marcus B. Finnegan Memorial Lecture stressed the benefits of universities working closely with industry:

I argue that universities as human social institutions, concerned as they must be with the well-being of all people, our own as well as those of the undeveloped nations of the world, must therefore necessarily be concerned, even more than they have been, with promoting the technological applications of science. Emphasis on such applied research, rewards for such research, greater funding for such research, especially in cooperation with corporations, may well bring forward many additional applications from universities that have not paid much attention to encouraging such efforts. The increased effort brings forward, also, a new set of problems.

If it happens that a university is successful in developing technology, how does it cope with success? It can deny it. It can ignore it. It can give it away. It can license it. It can promote it. It can exploit it. The sound policy, it seems to me, lies somewhere between denial and exploitation. I argue from three assumptions, themselves debatable.

First, universities should do all that is reasonably possible to earn returns on inventions, and should not be timid in making prudent business arrangements to assure the largest fair return.

Second, universities must provide an adequate reward to the inventor.

Third. a sizable part of the financial return to the university should be plowed back into research, especially the type of basic research that is difficult for industrial researchers to justify.[17]

During the past decades cooperative efforts between universities and corporations were somewhat restricted, looking to mutual benefits but generally excluding joint-venture research. Corporations provided management counsel to universities, accommodated cross-fertilization by permitting the executive on a "working sojourn" to teach and research in an academic community, faculty members on sabbatical leave spent time in corporate surroundings, and cooperative work-study programs were created for graduate and undergraduate students.[18]

Today it is evident that there is a serious gap between industry and the research resources of higher education. The results from university research projects are not being effectively conveyed to American industry on a preferential or priority basis. In addition, there is no efficient method for direct coupling of university research resources to industry for the performance of cooperative research. The research resources of this nation are primarily concentrated in universities and in private industry. Only the industrial sector has the capability to bring the benefits of research to the people.[19]

Barbara J. Culliton, writing for the *New England Journal of Medicine*, traces the growing reliance of universities on industry:

> If the 1980s witness a new growth of ties between the academy and industry, the reasons will come as no surprise. For more than a decade, federal support of basic research has been increasingly strained. There has been a real decline in the availability of federal research dollars, and the cost of doing the work of science has gone up. For some universities, the need to refurbish old buildings and to purchase new instruments for research and teaching has become acute. . . .
>
> A desire to lessen the entanglements that have come with federal dollars also lies behind the solicitation of industrial funds. The federal connection brought with it federal regulation. . . .
>
> An additional reason for science's turn to industrial sources of support has to do, not with regulation itself, but with the control over research that accompanies federal dollars. Academic investigators like to think of themselves as free to follow their scientific instincts when it comes to the design and execution of experiments. . . .
>
> Many factors have been involved in the academy's courtship of industry, but one factor seems overriding. For the first time in basic biomedical research, the university has something extremely valuable to sell.[20]

Reconciliation of the university's integrated triad of education, research, and service with corporate goals is the primary faculty objection to research relations with industry. Culliton states the question as follows: "Will the industrial connection warp the academic research enterprise?"[21]

In contrast to university operations, corporate management determines the new product objectives of a company and thus the research and development objectives. The problem confronting industry is how to make best use of whatever limited research and development capability a company possesses. These limitations probably cause new products to be less ambitious, imaginative, and farsighted. The first breath of life for innovative research and development originates with a creative scientist. Wise management should look for innovative ways to nurture these ideas to maturity. Such is the primary basis for industry–university cooperation.[22]

Organizational, administrative, and legal issues confronting the establishment of joint and cooperative research efforts between universities and industry are numerous. Some concerns are similar to those problems confronting university/government research efforts.[23]

Serious problems are inherent to industry–university cooperation, and they originate from fundamental differences between the two organizations. The following chart compares these generalizable differences in terms of publication of research findings, patents, trade secrets, the initiation and management of research, and the different time frames for conducting research. These differences, although not trivial, can be overcome through beneficial mutual negotiations.

In his statement before the Subcommittee on Science, Research, and Technology of the House Committee on Science and Technology, Donald Kennedy, President of Stanford University, stressed the immediate opportunities afforded higher education as a result of the commercializing of university biomedical research:

• A number of studies on the adoption of biomedical technology (such as, Julius H. Comroe, Jr. and R. D. Dripps, "Scientific Basis for the Support of Biomedical Science," *Science*, 192 [April 9, 1976], 105–111) illustrate that the time required for the diffusion of new innovations from the research laboratory to their application in the public interest is often disturbingly long—ten years or more. It can be argued that the involvement of commercial organizations in basic research will bring applications work closer to primary discovery, and thereby reduce the lagtime from conception to utilization in the stream of commerce.

• Commercial interest in basic research, whether it involves the attraction of scientists to industry or increased support of academic work by industry, will add a new and needed source of funding for such work at a time at which it is especially needed.[25]

Kennedy also identified some problems and pitfalls:

• Without additional provisions, there will be a strong tendency for the association of commercial interest with basic science to take the form it is now taking; that is, an increasing involvement of university faculty members in outside affiliations with particular companies. That prospect presents at least two difficulties. First there is the prospect of significant contamination of the university's basic research enterprise by the introduction of strong commercial motivations and potential conflicts of interest on the part of faculty members with respect to their obligations to the corporations in which they have consultancies or equity and their obligations to the university. Second, there is the danger of a sharp dissociation between research—as it moves more into industrial setting—and research training, which has always been a critical obligation of the universities and which industry is unequipped to do.

• The commercial environment is characterized by many more constraints upon

Comparative Table of University/Industry Basic Research Flow[24]

I. RESEARCH COMPETENCY

UNIVERSITY	INDUSTRY
Dependent on on-board faculty skills and laboratory facilities. We do not recruit personnel and build facilities to undertake new research ventures.	We maintain or acquire research competencies as needed. To enter a new field we hire skills or may acquire it through acquisition of a company already established in the field.

II. THE TIME CYCLE OF A RESEARCH PROGRAM

UNIVERSITY	INDUSTRY
Our ability to start a new project is affected by the academic calendar. Availability of graduate assistants is one factor. Availability of faculty usually depends on teaching responsibilities term-by-term.	Very flexible, and under company control, for start, stop, accelerate or phase out.

III. HOW RESEARCH PROGRAMS ORIGINATE

UNIVERSITY	INDUSTRY
A result of personal interest on the part of a faculty member or a group of faculty, leading to an outside sponsor. A few projects result from requests for proposals (RFP) from governmental or industrial sponsors which happen to coincide with research talents and interests of on-board faculty.	Projects originate from many sources. These include concepts of in-house researchers sold to management; management decisions to move into a new technical area; RFP's received governmental agencies or other companies. Starting a program is a management decision.

IV. MANAGEMENT OF RESEARCH

UNIVERSITY	INDUSTRY
Research is not managed in academe. The faculty principal investigator is responsible for administering the project, meeting sponsor requirements, budget control and reporting results. (All proposals to outside sponsors are reviewed by administration to ensure conformance with University and sponsor policies.)	Our research generally is highly applied or developmental in nature; basic research is rarely undertaken. We expect tangible and profitable results within a few years or we drop the program. Progress is reviewed carefully by management.

V. TRADE SECRECY

UNIVERSITY	INDUSTRY
No - "trade secrecy" is alien to the purposes of a university. From a practical standpoint, we cannot guarantee protection of proprietary information to a sponsor; at best we could offer a "best efforts" by faculty and graduate students.	Yes - trade secrets are a way of life, particularly in fast-moving high technology areas. Employees who divulge trade secrets can be and occasionally are prosecuted.

VI. PUBLICATION OF RESEARCH RESULTS

UNIVERSITY	INDUSTRY
Yes - our number one priority. Our reputation is dependent on publication in refereed journals.	Reluctantly, if at all. We publish only after making sure that it will not help the competition, i.e., damage our market position.

Comparative Table, *continued*

VII. PATENT PROTECTION

UNIVERSITY	INDUSTRY
A patent program is provided for three reasons: 1. A procedure for handling patents is required by sponsors of research programs—including all federal agencies. 2. Patents are an effective means of technology transfer, i.e., moving ideas from the bench to the public sector to benefit society. 3. Patents are a tangible recognition of faculty creativity—and (occasionally) they financially reward faculty.	We may or may not patent a new development. In highly competitive areas we often rely on trade secrecy as a means of getting a new product to market ahead of the competition. A patent is a publication—which makes it easier for the competition to counter the advance.
Policy: We expect to take title to all patents resulting from research.	Policy: In sponsoring research at a university, we expect to secure patent rights.

the openness and accessibility of scientific and technical information than is the university environment. Proprietary restraints on the free exchange of data have already begun to crop up at biomedical research meetings, and are presenting challenges to the policies of scientific societies and journals accustomed to open publication. Even more damage has been done to the informal roots of communication that characterize most vigorous fields of basic biological research.[26]

One solution proposed but rejected by both Harvard and Stanford is equity participation. Under this concept universities would go directly into business with their own faculties, thereby keeping the venture at home and avoiding risks of dissociating the research enterprise. Reasons for rejection are those articulated by President Derek C. Bok of Harvard in his 1981 report to the Board of Overseers:

The dangers of technology transfer to the quality of academic science are . . . four in number. First of all, the prospect of reaping financial rewards may subtly influence professors in choosing which problems they wish to investigate. . . .

The second concern is that professors may be diverted from *any* form of research (and teaching) in order to perform other tasks involved in the process of technological development. . . .

The third danger is the risk of introducing secrecy into the process of scientific research. Secrecy is, of course, anathema to scientific progress, since new discoveries must build upon what is already known. . . .

The fourth and final danger is a threat to the quality of leadership and ultimately to the state of morale within the scientific enterprise.[27]

Bok then expanded upon these ideas in his classic work *Beyond the Ivory Tower*. Academic science has always been fearful of the perverse law governing research that applied research invariably drives out pure research. Perhaps this attitude results from the traditional view of academic scientists which exhibits disdain for applied research. No university would support a situation in which academic investigators were influenced by powerful extraneous factors such as the prospect of large financial rewards. Universities will not support faculty neglecting their teaching and research in opting for massive consulting with corporations or lengthy experiments with entrepreneurship.[28] Paul Nash has stressed the need for academic researchers to have complete and uninhibited freedom to explore and investigate:

To the extent that external pressures and controls hinder the individual from thinking, speaking, reading, and writing freely, they threaten his healthy growth as a human being. Moreover, the need for such freedom, as Robert Calhoun has pointed out, "is greater, not less, as persons become masters of more extensive and complex ranges of experience."[29]

Bok notes that scientists depend on accurate and frank exchanges of their discoveries as reflected in their publications. Commercial motives can introduce a threatening form of secrecy. Universities must avoid the temptation which scientists may feel to withhold their discoveries until further developed into a patentable state in order to maintain a competitive advantage worth large sums of money. Universities must support the traditional ideal of science as a disinterested search for knowledge without ulterior motives. The continuance of disinterested inquiry helps to preserve the confidence and respect of the public in its institutions of higher education.[30]

Alternatives for universities are for them to regulate the participation of their own faculty members in outside ventures with industry. This option presents several difficulties. Industry consulting of various kinds has proven valuable in a number of other research sectors in the university, and longstanding policy and tradition permit it. The university may find itself suppressing valuable kinds of associations if it attempts to prevent the prospective injury associated with outside commitments by faculty. In addition, if universities overregulate faculty behavior it is possible that talented fac ulty members will end their university affiliations instead of those with the outside ventures. A 1962 survey indicated that thirteen percent of the faculty at four-year colleges and universities received outside consulting fees, whereas in a 1975 Ladd-Lipset survey the corresponding number was 48 percent. The same 1975 survey indicated that 89 percent of all faculty earned some supplemental income.[31]

Kennedy of Stanford stressed the danger of university control of faculty research and advocated a balance in research goals:

> [The] basic research system is characterized by a mixture of independent investigation and research training, and is characterized by an unusual open system of information exchange. For these reasons, it would seem to be in the national interest to keep the line between basic and applied research about where it is.[32]

Internal efforts to retain faculty and encourage their research efforts have led to adjustments which some would label as undesirable distortions of the university: is it desirable for research centers and laboratories to operate outside the structure of academic departments; does faculty consulting through interaction distract from their primary duties; is there a conflict of interest?

From industry's perspective additional difficulties manifest themselves. Corporate research interests are not well understood and cannot be inferred with accuracy from the items in existing product lines. No centralized entry point exists for the initiation of research proposals within most companies. Joint research with a university might threaten a research and development manager's internal operating budget since industry seldom appropriates funds for external research activities. Confidentiality of new technology is a major industrial problem as it seeks to claim exclusive proprietary rights which result from research and development. Licensing terms for proprietary rights to anticipated research results are extremely technical and require full-scale and time consuming negotiation.

Of mutual concern to both industry and the campus are other legal issues involving antitrust matters, copyright and patent questions, the tax questions applicable to the funds which industry "contributes" to research by a non-profit organization, product liability for defective or dangerous products, and liability and rights between industry and the university should the company be declared bankrupt.

In 1970, the trends in science affecting university research were recognized by J. R. Killian, Jr., then chairman of the Massachusetts Institute of Technology corporation:

> Today America is witnessing the opening up of new domains of science which can have profound effects upon society. This is a period of major biological discoveries which are yielding great advances in medicine and are unravelling the mysteries

of genetics. There is a growing collaboration . . . particularly between medicine and engineering. . . . Today universities, particularly their programs in science and engineering, are moving into a critical period. The government is cutting back on funds for research, and in a manner that is unplanned and therefore more damaging. This cut-back is made much more serious by steady inflation.[33]

A logical extension of Killian's growing collaboration in the sciences manifested itself in industry's growing interest in university scientific research. Both industry and the university viewed the corporate interest as an opportunity to replace government as a funding source. Wayne Biddle observed this phenomena as follows:

It would appear natural, then, that a working partnership might be born of these parallel difficulties. During the early 1970s, in fact, when the first slowdown of federal money for universities was exacerbated by cut-backs in aerospace research, and Department of Defense funding was stamped on by the Mansfield Amendment (which says that DoD campus money must be limited to projects with a "direct and apparent relationship to a specific military function or operation"), universities began to revive their ties with industry. . . . Corporate funding livened up considerably, needless to say, after inflation soared, resources began to seem finite, and the morally concerned classes graduated.[34]

After two decades of reliance on government research support, business and universities are renewing their alliance. President Bok of Harvard establishes five categories or guidelines that can result in a beneficial and positive link between industry and the scientific community in institutions of higher education. Advocating that the benefits of technology transfer with industry outweigh the risks, he encourages collaboration under the following headings:

[I] AVOIDING A DISDAIN FOR USEFUL RESEARCH It is often said that the highest goal of academic science is to pursue knowledge for its own sake and not for the purpose of achieving specific practical results. This ideal is constantly at risk in a world where scientific research depends on heavy support from public funds, for the public is chiefly interested in discovery not as an end in itself, but as a means to new products, new cures for disease, or new solutions to pressing social problems.[35]

There exists a traditional argument in academia that basic research is of greater intellectual significance than is applied (practical) research. Bok suggests that faculty who press this issue by deprecating the role of applied research could cause neglect of research problems of genuine intellectual challenge. Participation in joint research ventures has the benefit of allowing the scientist to investigate the problem with an eye toward eventual practical application for society's good. This kind of research will assist the graduate student in gaining experience for careers in industry and applied research.

[II] IMPROVING COMMUNICATION There are several opportunities to strengthen communication between universities and industry that carry little risk of impairing the quality of academic science.[36]

Mutual exchange mechanisms of various types exist to facilitate communication between industry and the university. An industry-associates program permits an information flow between business and academic science departments. Bok notes that consulting by individual scientists with industry is a common form of communication. By periodically visiting corporations, academic investigators exchange ideas, discuss recent developments, and share critical judgments that assist the process of technological development. This traditional consulting function has for years dispelled the theory that applied research is inferior and uninteresting and has allowed the university access to sophisticated research methods and new types of scientific instrumentation.

[III] OBTAINING PATENTS Patents offer universities an incentive to work harder to identify valuable ideas discovered in their laboratories. By making such efforts, they may speed the translation of knowledge into useful products and processes while eliciting new sources of revenue to strengthen academic science.[37]

Bok predicts that patent mechanisms will be refined by universities in the coming years. The development of patents by universities to a commercially feasible form will reduce obstacles which might prevent securing beneficial licensing arrangements with corporations. Corporations could then bring the fruits of academic science to the marketplace. Patents do not conflict with open access to knowledge. They only extend the use of information for commercial purposes by granting to the owner exclusive rights to a discovery for a period of seventeen years. Patent rights and the patent system were designed to encourage inventors to make known their discoveries instead of keeping them secret from competitors. Concerns that professors might be tempted to divert their normal research activities toward profitable patentable devices can be reduced by the university's sharing in the royalties with the professor. Contracting can ensure that the individual's rewards are not extremely large. Lastly, anxiety regarding the granting of exclusive licenses to the industry which supported a research project presents no difficulty. Such agreements are negotiated for a reasonable period of time to provide industry the opportunity for development and marketing. Should industry take no action by the end of the negotiated time period, then the license could be marketed to others.

[IV] CONTRACTING WITH PARTICULAR FIRMS . . . Thus one reason that many scientists prefer industrial funding today is that corporations usually offer their support with *less* control and *less* red tape than obtain under federal research agreements.[38]

Uneasiness over the potential problem that a contracting industry has access to information extending beyond the research agreement has no validity. Bok advocates that corporations do not purchase an entire school or department but simply create a working relationship with specified research scientists and their teams. The basic agreement between the partners normally will specify that publication of discoveries will not be restricted. Favored treatment by the university, however, should not be condemned or there would exist no incentive for a corporation to fund the research.

[V] INVESTING IN NEW VENTURES . . . If a university could own a substantial share of even one or two of these ventures, it might gain a source of added revenue that could go far to remedy the gaps and deficiencies in the current patterns of funding for research. . . .

Commercial ventures of this kind are bound to lead an administration into conflicts with several constituencies. For example, any effort to go into business with professors will expose the administration to almost certain disagreements, not only with its faculty partners but also with professors who feel envious or upset that their own cherished schemes have not received comparable support. Such commercial ventures will also impose upon the university unwelcome responsibilities it does not have as merely a tiny shareholder in a large established company. Investors who regard the university's participation as an endorsement of quality may feel aggrieved if it decides to pull out.[39]

The avoidance of the university investing in commercial ventures has many advantages. Faculty members remain important for traditional values and not as a source of some potential income to the institution. Academic administrators will not be caught up in overvaluing the work of their professors. Bok stresses that universities generally lack the business acumen to organize new business ventures the participation in which might result in political pressures and public relations problems for the non-profit university.

Cooperation between universities and industry is now rapidly becoming a significant force contributing to the innovation process. Indeed, such joint undertakings are so new that little opportunity has existed for the collection of such data or the actual comparison of contracts.

This text examines the use of complex research agreements for joint research activities between industry and universities as successful mechanisms to define and protect the legal rights of the contracting parties. These agreements can address various legal and productivity concerns thus resulting in positive performance benefiting both parties. These issues include industry's need for protection of patents and proprietary data and the faculty's desire for freedom to publish. Legal agreements in the form of negotiated contracts with adequate clauses articulating mutual considerations and obligations and which address potential problems have resulted in successful and beneficial research interactions.

Negotiations for cooperative research and technology transfer require participants to seek realistic accommodations of competing company and university objectives. For example, industry's need for privacy and security of its proprietary data and its need to obtain patent protection must be reconciled with the university's need to allow faculty researchers their freedom of association, expression, and publication. In addition, the university's desire for development of products for society must coexist with a company's need for freedom to modify its plans as dictated by changing market conditions, strategies, and objectives.

A cooperative research venture results in industry's generally receiving exclusive rights to new technology for commercial gain in exchange for the university's receiving financial support for its efforts and a potential share of commercial income. The university needs to recover its research costs, both direct and indirect, and must pass the expenses of patenting as well as product liability and patent litigation to its industrial partner to bear as normal business risks. The ultimate goal for the corporation and the incentive for the university is the potential reward through commercial success of the research venture.

The following chapters examine the historical development of these collaborative research agreements and the current legal issues being addressed in the form of negotiated contracts.

NOTES

1. Abraham Flexner, "The Graduate School in the United States," *Journal of Proceedings and Addresses of the Association of American Universities* (Chicago: n.p., 1931), pp. 114–115.

2. Letter from Thomas Jefferson, Monticello, to H. Lee, June 28, 1825, quoted in Richard J. Storr, *The Beginnings of Graduate Education in America* (Chicago: University of Chicago Press, 1953), p. 13.

3. *Encyclopaedia of the Social Sciences,* s.v. "Universities and Colleges," by Stephen d'Irsay.

4. Richard Hofstadter and C. Dewitt Hardy, *The Development and Scope of Higher Education in the United States* (New York: Columbia University Press, 1952), p. 70.

5. Don K. Price, "Federal Money and University Research," in *Science Policy and the University*, ed. Harold Orlans (Washington, D.C.: Brookings Institution, 1968), p. 23.

6. Hofstadter and Hardy, *The Development and Scope of Higher Education in the United States*, p. 88.

7. U.S. Congress, House, Committee on Science and Technology, *National Engineering and Science Policy, Hearings before a Subcommittee of the House Committee on Science and Technology*, 97th Cong., 1st sess., 1981, p. 29.

8. Ibid., p. 31.

9. Ibid.

10. Association of American Universities, *The Nation's Deteriorating University Research Facilities: A Survey of Recent Expenditures and Projected Needs in Fifteen Universities—Executive Summary* (Washington, D.C.: Association of American Universities, July 1981), p. 3.

11. *International Encyclopedia of the Social Sciences*, s.v. "Research and Development: Financing Social Science Research," by Henry W. Kiecker.

12. William F. May, "Corporate Support of Higher Education," in *The Corporation and the Campus,* ed. Robert H. Connery (New York: Praeger Publishers, 1970), p. 151.

13. Kim McDonald, "Science and Industry Discuss Research Partnerships at Universities," *Chronicle of Higher Education*, 5 January 1983, p. 4.

14. David F. Noble and Nancy E. Pfund, "The Plastic Tower: Business Goes Back to College," *Nation*, 20 September 1980, p. 246.

15. Robert M. Hutchins, *The Higher Learning in America* (New Haven, Conn.: Yale University Press, 1936 [2d printing 1962]), pp. x-xi.

16. Edward Schaffer, "The University in Service to Technocracy," *Educational Theory* 30 (Winter 1980):50–51.

17. Terry Sanford, "The University and Technology: New Paths and New Perspectives," in *The Law of Business and Licensing*, ed. Robert Goldscheider and Tom Arnold (New York: Clark Boardman Co., Ltd., 1982), pp. 1–66–1–67.

18. Clarke Wescoe, "The Use of University Resources," in *The Corporation and the Campus*, ed. Robert H. Connery (New York: Praeger Publishers, 1970), p. 56.

19. Edward L. MacCordy, "The Untapped University Resource," *Les Nouvelles* 13 (September 1978):178.

20. Barbara J. Culliton, "Biomedical Research Enters the Marketplace," *New England Journal of Medicine* 304 (14 May 1981):1197.

21. Ibid.

22. Edward L. MacCordy, "The Untapped University Resource," p. 180.

23. James Rowen, "Politics of University Research," in *Academic Super Markets,* ed. Philip G. Altbach (San Francisco: Jossey-Bass, 1971), p. 121.

24. U.S. Congress, House, Committee on Science and Technology, *National Engineering and Science Policy, Hearings before a Subcommittee of the House Committee on Science and Technology*, 97th Cong., lst sess., 1981, pp. 39–44.

25. U.S. Congress, House, Committee on Science and Technology, *Commercialization of Academic Biomedical Research, Hearings before a Subcommittee of the House Committee on Science and Technology*, 97th Cong., 1st sess., 1981, p. 14.

26. Ibid., pp. 15–16.

27. Derek C. Bok, "President's Report: Business and the Academy," *Harvard Magazine*, May-June 1981, pp. 29–30.

28. Derek Bok, *Beyond the Ivory Tower: Social Responsibility of the Modern University* (Cambridge, Mass.: Harvard University Press, 1982), p. 149.

29. Paul Nash, *Authority and Freedom in Education: An Introduction to the Philosophy of Education* (New York: John Wiley & Sons, 1966), p. 77.
30. Bok, *Beyond the Ivory Tower,* p. 149.
31. Robert H. Linnell, "Ethical and Economic Issues Facing Academe in a Rapidly Changing Society," in *Dollars and Scholars: An Inquiry into the Impact of Faculty Income Upon the Function and Future of the Academy*, ed. R. H. Linnell (Los Angeles: University of Southern California Press, 1982), p. 3.
32. Kennedy, in U.S. Congress, House, Committee on Science and Technology, *Commercialization of Academic Biomedical Research, Hearings before a Subcommittee of the House Committee on Science and Technology*, 97th Cong., 1st sess., 1981, p. 18.
33. J. R. Killian, Jr., "University Research," in *The Corporation and the Campus*, ed. Robert H. Connery (New York: Praeger Publishers, 1970), p. 41.
34. Wayne Biddle, "A Patent on Knowledge: Harvard Goes Public," *Harper's*, July 1981, pp. 22–23.
35. Bok, *Beyond the Ivory Tower*, pp. 151–152.
36. Ibid., p. 153.
37. Ibid., p. 154.
38. Ibid., pp. 158–159.
39. Ibid., pp. 160–161.

2

The Historical Development and Legal Foundation of Industry–University Collaboration

BASIC AND APPLIED RESEARCH IN THE UNIVERSITY

The primary functions of American academic institutions include teaching, research, and service to society. University research has traditionally been oriented in the direction of the educational experience and the extension of fundamental knowledge. Research emphasis has historically been toward basic science and technology, not development and commercialization of products. This research dichotomy reflects a traditional tension in higher education between the theoretical and practical. This was recognized by the late W. H. Cowley, David Jacks Professor of Higher Education at Stanford University, when he described his logocentric/practocentric/democentric theory.[1]

As stated by President A. Bartlett Giamatti of Yale University, the definition of basic (or fundamental) research in the academic environment involves the quest for new knowledge on the part of the researcher with no commercial goals as the end product:

> Basic research, that is, investigation that seeks new knowledge and understanding rather than solutions to immediate problems, is the essential nature of research on the part of all scholars. It obviously includes but is not restricted to basic research in the biological, medical, physical and many social sciences.[2]

Applied research has been described by then Chancellor Clifford C. Furnas of the University of Buffalo as an undertaking carried on for exploitation or application in commercial or industrial fields, whereas basic research may be arbitrarily defined as undirected research pursued solely for satisfying human curiosity and with no direct practical application intended or in mind.[3] This definition places emphasis on the service function of the university.

Clark Kerr recognized two major forces that have molded the modern American research institution, both of which came primarily from the federal government:

The first was the land grant movement. . . . This act set the tone for the development of American universities, both public and private, for most of the ensuing hundred years. It was one of the most seminal pieces of legislation ever enacted. . . .

Supporting the impact of the land grant movement was the effect on American universities of the model supplied by Germany. This German model gave academic respectability and content to the "land Grant" idea. . . .

The second great impact on the universities began with federal support of scientific research during World War II. The war time laboratories that were the forerunners of such continuing government-financed research centers as the Lincoln Laboratory at the Massachusetts Institute of Technology, The Argonne Laboratory at Chicago, and the Lawrence Radiation Laboratory at California, opened a new age. The major universities were enlisted in national defense and in scientific and technological development as never before.[4]

Connections to the acceptance of basic research as a major goal of universities were recognized by the Land Grant College Act (Morrill Act) of 1862.[5] This remarkable assistance to state and private education through the federal government's sale of land created a fruitful relationship between an important university sector and American agriculture and industry. As Earle D. Ross states:

The purpose was sufficiently broad and general to include all the varied concepts of industrialists: ". . . the endowment, support, and maintenance of at least one college where the leading object shall be . . . to teach such branches of learning as are related to agriculture and the mechanic arts . . . in order to promote the liberal and practical education of the industrial classes in the several pursuits and professions of life."[6]

The cooperative keystone of the land-grant formula lies in the joint founding and subsidy by both Federal and state governments. The control of education traditionally falls within a state's prerogative as this nation's constitution makes no mention of this area as one of Federal involvement. Morrill asserted in reply to President Buchanan's veto of the original bill for constitutional reasons that his proposed act left the entire control in the hands of the individual states to arrange and manage as they saw fit.[7]

Subsequent Federal activity has been justified by national needs in both agriculture and industry. Research and university sponsored extension services involved regional and national, rather than purely state, concerns. Federal interest was necessary to insure the continuance of fundamental research; state support might lead only to work on immediate practical problems, the solution of which would be impossible without continued basic research.[8] Through the Morrill Act we find the first sponsored efforts toward university involvement in both basic and applied research.

GOVERNMENTAL RESEARCH SUPPORT—NINETEENTH THROUGH TWENTIETH CENTURY

The Federal government has encouraged and supported science and has profited from it since the founding of this nation. Early government financial support provided a significant source of strength to scientific development in America. Congressional activity through statutes emphasized exploring, surveying, patenting, agriculture, and technological developments in the nineteenth century. Federal interests centered around support for basic research.[9] Notable examples included the research efforts of the Department of Agriculture (1862, 1889),[10] the creation of the Coast Survey (1807, 1832), the National Institute for the Promotion of Science (1840), the founding of the Smithsonian Institution (1846), the American Association for the Advancement of Science (1848), and the National Academy of Sciences (1863).

The twentieth century saw this interest in applied research expanding through the creation of specialized research institutions which conducted essential scientific work: The National Bureau of Standards was created in 1901, the Public Health Service in 1912, and the National Advisory Committee on Aeronautics in 1915. Agricultural research services including significant university support illustrated unique and flexible characteristics, accelerating America's leadership in agriculture. During World War I an impressive federal scientific establishment took shape to support the research needs of the Army, Navy, and civilian agencies.[11]

The Council of National Defense was created in 1916 to coordinate industrial use of resources for the national security and welfare.[12] The Council was authorized to organize subordinate bodies for its assistance in special investigations, including the creation of committees of specially qualified persons who could advise in research efforts. Through spinoffs and mutations the War Industries Board and Food Administration emerged under the leadership of Bernard Baruch and Herbert Hoover to control large segments of the economy. The War Industries Board developed a technical and consulting staff with the Mellon Institute in Pittsburgh.[13] This historical growth of government sponsored research did have some serious faults. Four major sectors of scientific support—government, universities, industry, and private foundations—had developed along separate and individual lines. During the 1930s, their activities were not integrated; they lacked effective centralized leadership and coordination and showed inadequacies in communication among the different segments of the system.[14]

Bruce L. R. Smith and Joseph J. Karlesky noted in their classic study of academic science the decline of federal government involvement in in-house laboratories during the 1930s.

Depression-era budget cuts caused many bureaus to lose their best scientists to other centers of research. Although the Department of Agriculture, with its experiment stations, and land-grant colleges, maintained its high level of R & D, most other federal departments, including the military, were forced to cut back. In contrast, American universities, with assistance of federal grant programs to the states, enhanced their positions as performers of both basic and applied research—most notably, in the case of applied research, with respect to agriculture.[15]

Upon the conclusion of hostilities at the end of World War I, a significant increase occurred in the number of industrial research laboratories as businesses discovered science as a tool for product development, testing, innovation, and diversification. Indeed, research proved essential in certain growth industries such as specialty chemicals, pharmaceuticals, and electrical manufacturing. Private philanthropic support for science from the Rockefeller and Carnegie Foundations was directly limited by the Depression. Rising research costs resulted in the exclusion of foundations as a major supporter of scientific research although they renewed significant special program areas such as manpower research and training.

A peacetime role for continued Federal involvement in research was attempted by President Woodrow Wilson's Executive Order of May 11, 1918,[16] creating a permanent National Research Council (deriving Congressional sanction from the National Academy's Act of 1863) to advise the government; to stimulate research in mathematical, physical, and biological sciences; to formulate comprehensive projects of research; and to direct attention of scientific and technical investigators to important military and industrial problems in the interest of national defense.[17] This agency struggled for existence on the signing of the Armistice. The peacetime National Research Council (NRC) relied on money exclusively from foundations including the Rockefeller and Carnegie Foundations. The postwar NRC limited its operations to a scale set by its private funding and did effective work in assisting U.S. participation in international scientific experiments. It eventually lost touch with government, resulting in an end to its role as coordinator and advisor to government bureaus. Other agencies of governmental research such as the National Advisory Committee on Aeronautics and the Public Health Service went into a decline as a result of Congressional funding reductions.[18]

The economic depression that followed the 1929 stock market crash had a serious impact on American science. The national interest in research by government, industry, and universities was so interconnected both internally with each other and with the total economy that the disruptions of the depression affected the whole research structure. President Roosevelt began in 1935 to create a central scientific organization to benefit both the government and the country. By June of 1935 the National Resources Committee became active in salvaging, planning, and coordinating research efforts through its science committee. Even the Works Projects Administration developed projects that provided professional, technical, and manual workers to assist in scientific and technological research and experimentation at tax supported universities and colleges.

In his definitive study of the federal government and its role in science, A. Hunter Dupree observed:

> The decade 1929–1939 . . . was one of increasing use of research in the government. Much energy during the decade went to combating the effects of the depression. But this effort itself combined with the cascading scientific discoveries to give government a more stable and flexible establishment. Viewed as the last stepping stone to World War II, the decade of the depression and New Deal made definite contributions. The concept of research as a national resource served well in the crisis. The techniques such as the use of contracts for joining federal and nonfederal research were valuable in setting up an emergency organization.[19]

In summary, the growth of scientific support since the 1930s was highlighted by expanding interrelations within government and education. The federal government with its national defense interests featured as the paramount coordinator with a central role in this movement. In basic research, the key to this development became the partnership between government and universities. World War II finally forged the shape and character of government–industry–university relations. Because the limited research programs of the armed services proved insufficient to meet priority wartime demands, the government sought major assistance from university scientists.

This new framework for government–industry–university research relations was established in 1941 when President Roosevelt created the National Defense Research Committee (NDRC) and a year later expanded it into the Office of Scientific Research and Development (OSRD).[20]

The Office of Scientific Research and Development operated no laboratories but contracted with individual industries and universities. Through contracting, OSRD was able to dedicate the whole institutional laboratory to government service and in this process important precedents were established for future research relations between the government and universities in support of scientific research.[21] Under OSRD direction contracts set cost standards and articulated the responsibilities between the government's scientists and budget officers and the university's principal investigators and administrators. No differentiation was made between the private and public university. Patent guidelines were stated as well as rules for maintaining security and directing property disposal. Research was permitted by individual researchers on small scale assignments, and large, qualified institutions were awarded large scale projects.[22]

With the conclusion of hostilities in 1945 OSRD went out of existence. However, the foundation was laid for a national science support system. President Roosevelt's recognition of the need for the federal government to maintain an active role in shaping future national science policy resulted in the President's empowering Vannevar Bush, former director of OSRD, to assemble an advisory committee representing the scientific community to recommend future science support policy. The Bush Report helped establish the tone for postwar research relations between government and science. As this report notes, the government was seen as a central supporter and promoter of research capabilities:

The most important ways in which the Government can promote industrial research are to increase the flow of new scientific knowledge through support of basic research, and to aid in the development of scientific talent. In addition, the Government should provide suitable incentives to industry to conduct research, (*a*) by clarification of present uncertainties in the Internal Revenue Code in regard to the deductibility of research and development expenditures as current charges against net income, and (*b*) by strengthening the patent system so as to eliminate uncertainties which now bear heavily on small industries and so as to prevent abuses which reflect discredit upon a basically sound system. In addition, ways should be found to cause the benefits of basic research to reach industries which do not now utilize new scientific research.

The Government should accept new responsibilities for promoting the flow of new scientific knowledge and the development of scientific talents in our youth. These responsibilities are the proper concern of the Government, for they vitally affect our health, our jobs, and our national security. It is in keeping also with basic United States policy that the Government should foster the opening of new frontiers and this is the modern way to do it. For many years the Government has wisely supported research in the agriculture colleges and the benefits have been great. The time has come when such support should be extended to other fields.[23]

The Bush Report was placed into legislative form in Senator Magnuson's bill to establish a National Research Foundation. Simultaneously, in 1945, Senator Kilgore's Subcommittee on War Mobilization of the Committee on Military Affairs issued reports recommending what type of governmental agency should be established to continue the high level of scientific research necessary for ongoing national security purposes. This Subcommittee recommended a peacetime National Science Foundation.[24] A key difference between the Bush approach and the Kilgore recommendation rests in the Foundation's organization. Don K. Price of Harvard, who served on the Commission on the Organization of the Executive Branch of 1948, pointed out:

The Kilgore bill provided that the Foundation be headed by an administration appointed by the President, who would have the benefit of advice of a part-time Board, but would not be bound by such advice. This was in sharp contrast with Dr. Bush's proposal that the Foundation be headed by a part-time Board that would elect its own executive.[25]

Thus, the continued involvement of government in the national scientific research effort was assured, especially in controlling the distribution of federal funds.

Other government agencies quickly created research branches to fill gaps left by OSRD. The creation of the Office of Naval Research in 1946, followed by similar offices in other services, reflected the military's desire to continue its relationship with basic scientific research. The National Institutes of Health, although founded in 1930, began a vigorous extramural grant program for general medical research. The postwar era saw other agencies such as the Atomic Energy Commission beginning to award project grants and contracts to university scientists. With the university-run laboratories of the Manhattan Project as a foundation, the Atomic Energy Commission created a system of national laboratories that became scientifically preeminent in many fields.

Despite these limited research achievements by the agencies, the government still did not fully and explicitly articulate its framework for postwar science affairs. It had not clarified the role of university-based science as a significant national resource. Industrial involvement in the national scientific research effort remained unclear. Thus, in 1947 President Truman appointed a President's Scientific Research Board chaired by John R. Steelman to address these and other research issues. Significantly among its recommendations the Board recognized in its report of 27 August 1947 the future role of the university in governmental science policy:

1.That, as a Nation, we increase our annual expenditures for research and development as rapidly as we can expand facilities and increase trained manpower. By 1957, we should be devoting at least 1 percent of our national income to research and development in the universities, industry, and government.

2.That heavier emphasis be placed on basic research and development budgets. Expenditures for basic research should be quadrupled and those for health and medical research tripled in the next decade, while total research and development expenditures should be doubled.

3.That the federal government support basic research in the universities and nonprofit research institutions at a progressively increasing rate, reaching an annual expenditure of at least $250 million by 1957.[26]

Several keystone decisions that affected postwar support by the government of academic research were made a few years after World War II. That basic pattern has been stated under the following broad principles:

First, to assure that scientific research would further the broad aims and priorities of government as defined by the Congress and the President, federal research support would be closely linked to the operating missions of federal agencies; smaller shares of funds would be devoted to basic sciences and to the research-related educational functions of universities;

Second, non-government laboratories would be relied upon to perform much of the research of interest to the federal government;

Third, direct relationships between federal agencies and university researchers were established;

Fourth, university research and graduate training were viewed as closely related functions, thus avoiding the pattern in some nations of separating advanced scientific training from research activities.[27]

No real definitive history exists covering national research efforts during the 1950s and 1960s.[28] Daniel S. Greenberg in his investigative text *The Politics of Pure Science* examines the political maneuverings between scientists and government between the end of World War II and 1967. Noting that World War II had established the romance of science and government, Greenberg saw this as rapidly ending in 1957 with President Eisenhower's recurrent economy drives. Secretary of Defense Charles E. Wilson in the Eisenhower administration began the steps for the wholesale reduction of the Defense Department's scientific support:

Wilson announced a cut of 10 percent in Defense support for basic research, though it was estimated that, because of lagging budgets and inflation, the Department, in 1957, was actually supporting 25 percent less basic research than it had in 1952. . . . [T]he Department's scientific clients were energized to a state of panic by Wilson's plans. The Air Force Office of Scientific Research was on the verge of dispatching telegrams canceling 600 research projects, most of them basic research projects in universities, when it tried a last desperation move. It appealed to . . . the executive branch's highest ranking science advisory body, the Science Advisory Committee of the Office of Defense Mobilization, took the cause to the White House, and returned with reports of a sympathetic response. But the appeal was almost immediately rendered unnecessary. In October 1957, the Soviets orbited Sputnik, and honeymoon fervor immediately returned to the romance between science and government.[29]

The awakening to the reality of Soviet superiority in science and technology had an overwhelming effect on improving political support for scientific research funding. A new phase in federal relations with the intellectual community returned in 1958 with the passage of the National Defense Education Act (P.L. 85–864),[30] which provided extensive federal aid to education. The politically unifying effect of the Soviet space effort is observed by Daniel S. Greenberg:

Sputnik, in a moment, sent the pendulum swinging back. Not only did it inspire the release of a flow of funds, but it also inspired government to institutionalize many parts of the science advisory system that had randomly sprung up throughout Washington after the war. The Science Advisory Committee, buried within the obscure Office of Defense Mobilization, was elevated to the status of the President's

Science Advisory Committee (PSAC). . . . Thus, courtesy of the Soviet space program, the science–government relationship resumed honeymoon status.[31]

Although Sputnik had a powerful impact on increasing governmental support for research, it failed to maintain its momentum. By the early 1960s, reaction to budgetary reductions had again returned. Beyond budgetary issues, Congress by 1963 began to probe into why funds were sought, where they were to be spent, under what circumstances, and with what assurances of accountability. Science was viewed as one with technology. The concern of Congress underscored the belief that technology had become too expensive, too poorly understood in terms of economic values to be left to mere scientists. Basic research support had become another government expenditure to be examined, criticized, and reduced.

Yet support for higher education through the National Defense Education Act of 1958 continued as subsequent Congresses amended and expanded the original law. Barbara B. Clowse of the University of North Carolina examined this phenomenon and its impact on research at graduate levels in her text on this statute:

By 1968 those who administered the defense–education law could look back on quite a record. They had spent $3 billion on a multitude of projects. As many as fifteen thousand NDEA fellows had been supported in graduate school and 1.5 million undergraduates had been assisted with loans. . . .

The legislative pattern of merging educational-aid programs continued throughout the succeeding decade. In 1972, after nearly two years of consideration, Congress enacted the Education Amendments Act. Legislators gathered into one title of this behemoth most federal commitments to higher education. . . .

Although any association of the Sputnik crisis with these educational programs has long vanished, it would be well to remember its importance as a catalyst.[32]

Examination of the Executive Branch and its involvement in the formulation of science and technology policy from the post–World War II time period into the Carter Administration is discussed by James E. Katz in his study *Presidential Politics and Science Policy*.[33] Written primarily to analyze the role of individual presidents in creating science policy as it relates to the power of the presidency, this text provides an excellent historical overview of government's involvement in advancing scientific research. Katz notes that the president plays a significant role in our national science policy:

Science policy decisions are made on a political basis and the White House has played, and continues to play, a major role. . . . [T]he president and other politicians impinge on the internal mechanisms of scientific research per se, rather than the social organization of the scientific community. This effect reveals itself not only in terms of broad national priorities for research and various large-scale campaigns such as the "War on Cancer" or "Landing on the Moon," but also on less-publicized areas of policy, and finally on the individual project level as well. Presidential priorities directly influence the training and supply of scientific manpower, who goes where to be trained, what types of jobs will be available, and, to an awesome extent, even what the research priorities are in a given discipline.[34]

Claude E. Barfield, while a visiting fellow at the American Enterprise Institute, authored his text *Science Policy from Ford to Reagan*. This study examines the research and development strategies of the Ford, Carter, and Reagan administrations through an examination of the policies of the three administrations. Barfield recommends that government's primary concern should be directed toward long-term, basic research elements in the research and development budget. He notes the following reasons why the private sector will not support basic research at an adequate funding level:

In the future, the primary attention and concern of the White House and Congress should be redirected toward the long-term, basic research elements of the federal research and development budget. There should be less attention and concern focused on swings in the applied end of the R & D spectrum. . . . This proposal logically follows from an agreement that has developed over several decades . . . that the private sector will not support basic research at an adequate level. It will not do so because individual firms cannot retain many of the benefits from the research; because of the risk and uncertainty of the timing of the benefits; and in some cases because there is the need for huge, long-term investment.[35]

Edwin Mansfield and his associates at the Wharton School of the University of Pennsylvania recorded the process of technological change and its importance to American corporations from an economic viewpoint in a series of texts covering 1968 through 1982.These studies look closely at the organizational and strategic factors in American industry that account for successful research, development, and innovation.[36]

Other literature examines research and development by government, industry, and universities covering the time period of the early 1960s (and in one instance from 1953) to date and reflecting the statistical data collected by the National Science Foundation in a variety of studies.[37]

National expenditures for research and development between 1953 and 1967, in 1976 dollars, grew by more than 350 percent. During that time period the federal government increased its own research and development expenditures by almost 425 percent. By 1967, the federal government was providing 62 percent of the national research and development investment. However, since 1967, the growth rate has been far below that of the previous decade. By 1976 the federal contribution to the total had declined to an estimated 53 percent.[38]

With reference to the financial support for research in constant dollars, the 1953 to 1977 growth appears to proceed at a very accelerated rate.

However, the post–1970 inflation rates rose significantly, and once the dollar figure is converted the picture changes. If viewed in constant dollars, the national research and development support in 1977 becomes roughly 2 percent above that in 1967.The federal portion declines by almost 1.7 percent. The National Science Foundation expected non-federal research and development spending to rise slightly in 1977, with the largest percentage increases occurring in universities and colleges and non-profit institutional sectors by 9 percent and 10 percent respectively.[39]

In reviewing the weakening of industrial interest in basic research over the post–World War II period to the early 1960s, the National Science Board identified various reasons for this trend:

> First, management held overly optimistic expectations of the role of and possible returns from very broadly conceived basic research;
>
> Second, the management of laboratories and their integration into the rest of company activities left much to be desired;
>
> Third, there appeared a shift in balance at top corporate levels from managers with operational experience to those with a mere cost-accounting approach to management who placed less emphasis on long-range activity necessary for fundamental research;
>
> Fourth, recessions in the late 1950's and early 1970's caused many corporations to reduce their research budgets.[40]

Recent data on industrial research and development support at institutions of higher education suggest that only after 1970 did the percentage of industrial support of university research and development rise above 3 percent reaching a 1981 level of 3.8 percent. In constant 1972 dollars industrial support for academic research doubled between 1966 and 1978.The data for 1980 indicate that, of the 200 campuses reporting, only 25 universities have more than 10 percent of their research and development expenditures coming from industry. General conclusions from this National Science Foundation data point to the fact that industrial support for university research is increasing, that the amount of funding available from private industry has in some instances filled gaps left by reduced federal expenditures, and that large payments of industrial funding usually are concentrated with only a select group of prestigious universities having strong research reputations. Industrial support of academic research for FY 1980 is estimated at $235 million in current dollars.[41]

The magnitude of university research supported by industrial contracts as opposed to research supported by corporate philanthropy has never been clearly investigated. A 1982 survey of over 200 corporate chief executive officers reported that 60 percent of the corporations anticipated increasing their charitable contributions to higher education beyond the inflation rate over the next few years.[42] Existing data on corporate philanthropy to higher education caused the National Science Foundation to estimate reasonably that academic research supported by corporate gifts and grants roughly equals that supported by corporate contracts.[43]

THE ADVENT OF INDUSTRY–UNIVERSITY RESEARCH AGREEMENTS

Industry's involvement in university science and engineering began in the late nineteenth century and has increased slowly in intensity as documented by David F. Noble in his text *America by Design: Science, Technology and the Rise of Corporate Capitalism*. Noble observes that there have been ups and downs in the industry–university relationship with long, flat periods after World War II when government's generosity for funding scientific research precluded businesses from competing. He notes that there exists an increasingly important role of the university in aiding corporate research:

While the primary mission of the university within the industrial system was the "efficient production of human material" according to "industrial specifications"—which made not only the building of universities, but education itself an industry—the role of universities as centers of research for industry was also a vital one. After its first survey of research activities in the United States, the National Research Council maintained that "the main sources of research in America have been, and must continue to be, the universities. . . . "

Industrial sponsorship and direction of university-based scientific research successfully shifted the burden of some significant costs, and risks, of modern industry from the private to the public sector. But this was not all. Perhaps more important, it redefined the form and content of scientific research itself. This involved more than the general shift away from the search for truth and toward utility which had already been well underway by the turn of the century. Now the shift toward utility assumed particular forms, molded by the specific, historical needs of private industry, by particular firms intent upon increasing their profit margins and their power. This reorientation affected not only what kinds of questions would be asked but also what particular questions would be asked, which problems would be investigated, what sorts of solutions would be sought, what conclusions would be drawn. Science had, indeed, been pressed into the service of capital.[44]

University and industry linkages go back as far as the founding of Massachusetts Institute of Technology (MIT) in 1861. MIT has had the explicit mission to establish close research ties with industrial interests. Gilbert S. Omenn, Chairman of the Department of Environmental Health at the University of Washington School of Medicine, commented on this early relationship:

MIT started an Industrial Associates Program after World War I which now involves more than 200 companies. For a fee, the company gains a window in the state of the art in the academic laboratories and the faculty, staff, and students gain access to a wide variety of corporate R & D problems.[45]

Omenn also notes the historically significant contribution to cooperative technology transfer of the Wisconsin Alumni Research Foundation (WARF):

Another model for cooperative technology transfer arrangements is the organized patenting and licensing approach. Every discussion of these university efforts begins with the Wisconsin Alumni Research Foundation. The Wisconsin Alumni Research Foundation (WARF) was chartered in 1925 at a time of considerable ferment at the University. Wisconsin was already the nation's leading institution in number of doctorates awarded in the sciences. . . .

. . . A group of alumni founded WARF, filed patent applications on behalf of Professor Harry Steenbock for the methods he developed for irradiating milk to activate vitamin D—when neither the university nor the attorney general wanted to take the risks or incur the costs of patent applications—and established an arm's-length relationship which has returned over $100 million to the university for research grants, professorships, traveling fellowships, and student support in a host of departments throughout the university.[46]

Publicity involving contemporary industry–university research coupling began to intensify around 1978. Although authoritative statistics regarding levels of support are largely anecdotal since 1980–81, evidence in scientific journals and the higher education media suggest strong evidence of increased activity. A variety of financially large and visible agreements for long term research collaboration between companies and universities has been consummated within the past few years: Harvard-Monsanto; Washington University–Mallinckrodt, Inc.; Harvard Medical School–Seagrams; MIT–Exxon; Carnegie Mellon–Westinghouse; Massachusetts General Hospital (Harvard)–Hoechst A.G.; and Washington University–Monsanto.[47]

Incentives for enhancing the university–industry connection are significant. From the university's perspective, traditional and increasing research interests are involved in solving critical domestic problems and a renewed appreciation for the role of industry in such problem solving. The practical need of academia motivates the university to find alternative funding to replace increasingly scarce funds from its major research sponsor—the federal government. Additionally, government regulations related to scientific and financial accountability, human and animal experimentation, biohazards, and affirmative action have arguably reduced the efficiency, flexibility, and independence of the academic scientist. Lastly, declining student enrollments and decreasing employment opportunities in academic and government research centers have forced university administrators and faculty to regard industry as a source of research support and as a potential employer of its graduates. Other related linkings include utilizing industry for part-time faculty and as a consumer of major continuing education programs.

> Perceived benefits for universities include the potential for long-term research support less entangled in red tape; help from industry in making new knowledge and technology commercially useful; broader educational experience, industrial exposure, dissertation topics, and potential employment opportunities for students, and stimulation of university faculty through interaction with industrial scientists and engineers and through access to specialized industrial equipment. . . .
>
> Potential benefits for industry include additional sources of ideas, knowledge, and technology on which to base potential new products and processes; ability to draw upon competent scientists from around the country without expanding in-house capabilities; high benefit to cost ratio when compared to building an in-house research unit; source of potential research employees sympathetic to industrial needs; and stimulation of industrial scientists and engineers.[48]

The types of relationships that have been created have been generally categorized by the format of the research agreement—simple contracting for a single purpose, cooperative research programs, consortia, industrial park leasing arrangements, limited partnerships, joint ventures. Relationships range from undirected financial contributions to a significant ongoing research partnership. The future will unfold a whole spectrum of industry–university interactions depending on the goals and objectives of each organization and its institutional characteristics. Relevant factors identified by Denis J. Praeger and Omenn include:

> • the size, structure, and profitability of the industry, the nature of its business, and progressiveness of its research program;
>
> • the type, size, and financial health of the university, the relative size and stature of its science and engineering programs, and the orientation of its research and researchers;
>
> • the influence of external factors such as geographic proximity, location of university alumni in key industry positions, and migration of faculty to industry may be influential.[49]

A variety of studies have identified multiple types of industry-university relationships which have developed over several decades. These include:

> Corporate contributions to the university in the form of unrestricted gifts, capital contributions to departments, centers, or laboratories, and fellowships for graduate students;
>
> Mutual procurement of services by industries and universities using students for on the job training or thesis topics or educating and training employees; cooperative research projects of mutual interest; research consortia;

Reseach partnerships involving long-term research programs mutually beneficial, usually involving contractual arrangement;

Patent licensing by the university to industry on an exclusive or nonexclusive basis;

Entrepreneurial role by university as a managerial partner or equity holder in conjunction with industry or others.[50]

Edward E. David, Jr., President of Exxon Research and Engineering Company, urges more attention on agreements between academia and industry supporting fundamental research. Powerful influences such as intense economic and military competition, skyrocketing research costs, and a need to reduce time from inception to the market place are forcing researchers from two diverse areas to work together. Agreements will necessarily take a variety of forms reflecting the extraordinary differences of the nation's industrial and academic institutions. Academia, industry, and the nation will benefit as academic research reinforces the industrial innovation system in a new partnership.[51]

Higher education recognizes the need to renew the links with industry in research cooperative efforts as part of the technological imperative facing universities. Many authorities recognize the necessity for industry and higher education to form intimate ties for their mutual benefit.[52] Higher education is adaptable and corporate support through research agreements will assist in reducing the campus controversy. The university president and corporate executive recognize that the development of ground rules and guidelines for industry–university relationships through contractual agreements need to exist and can be successfully articulated.[53]

FEDERAL INTEREST IN INDUSTRY–UNIVERSITY RESEARCH AGREEMENTS

The federal government's traditional interest in the industry-university venture continues as part of its renewed policy interest in the role of research and development in economic growth and in the role of research as the foundation of new technology. Since the early 1970s, the executive branch supported industry–university cooperation as illustrated by President Nixon's message to Congress on 16 March 1972. He recommended action by both the National Science Foundation and the National Bureau of Standards to

> . . . determine effective ways of stimulating non-Federal investment in research and development and of improving the application of research and development results. The experiments to be set up under this program are designed to test a variety of partnership arrangements among the various levels of government, private firms, and universities. They would include the exploration of new arrangements for cost-sharing, patent licensing, and research support, as well as the testing of incentives for industrial research associations.[54]

This resulted in the National Science Foundation's creating experimental programs involving industry–university interaction. The most significant of these included: the University–Industry Cooperative Research Centers Experiments (1973), the Innovation Centers Experiments (1973), and the University–Industry Cooperative Research Projects Program (1978).[55]

Concurrently in 1973 the National Bureau of Standards launched the Experimental Technology Incentives Program (ETIP) to conduct policy studies and experiments in cooperation with government agencies having primary policy responsibility. The ultimate goal lay in the assistance to the participating agency in formulating a new policy together with testing it through a formal, evaluative experiment. Several examples involved direct participation of universities, and thus the policy experiment can be classified as innovation in industry policy research. Policy experiments included the patenting of university research and the joint adoption by a university and an adjoining urban community of an advanced energy supply system.[56]

In 1978 the executive branch initiated a domestic policy review of industrial innovation, and industry–university cooperative research efforts formed a significant segment to be reviewed. During 1979, the Carter administration made several recommendations to expand existing Federal industry–university programs and to expand these programs to other agencies. These proposals were incorporated into the Stevenson-Wydler Technology Innovation Act of 1980 (P.L. 96–480)[57] and are generally reported in House of Representatives hearings *Government and Innovation: University–Industry Relations*.[58]

Additionally, the 96th Congress passed further legislation encouraging the commercialization of innovative ideas and technologies developed as a result of Federal sponsorship of university- and industry-based research and development projects. The Uniform Federal Patent Policy Act of 1980 (P.L. 96–517) permits universities, non-profit firms, and small businesses to elect to take title to inventions resulting from Federally funded research and development activities.[59]

In 1980, the Antitrust Division of the U.S. Department of Justice issued its *Antitrust Guide Concerning Research Joint Ventures*.[60] This important document was intended to clarify for private industry the conditions and guidelines under which cooperative research and other research and development efforts with universities would be permitted under the antitrust laws.

With the advent of the Reagan administration in January 1981, there was a significant shift in the concept that the Federal government's role in industry–university relationships should be one to stimulate research and development, especially in regard to industry–university research interactions. The major emphasis of the Reagan policy is directed toward providing incentives for research and development investments through tax legislation. The Economic Recovery Tax Act of 1981 (P.L. 97–34)[61] includes a variety of provisions aimed at encouraging accelerated support for industrial research and development with this general concept including interactions with universities.

In addition to the legislative branch's hearings mentioned above, Congressional interest remains active as reflected by various hearings held over the past several years.

1) 24 March 1980—Hearing before the Subcommittee on Science, Research, and Technology of the Committee on Science and Technology (U.S. House. 96th Congress, 2d session).

The Government-University Accountability Relationship in the Field of Scientific Research.[62]

This hearing examines the extensive review conducted by the National Commission on Research with a primary focus on the Government-university partnership, its problems and possible solution especially regarding issues of accountability primarily under the Federal Grant and Cooperative Agreement Act of 1977 (P.L. 95–224).[63]

2) 8–9 June 1981—Hearings before a Subcommittee on Science, Research, and Technology of the Committee on Science and Technology (U.S. House. 97th Cong., 1st session).

Commercialization of Academic Biomedical Research.[64]

This investigative hearing by Hon. Albert Gore, Jr., focuses on current ethical and institutional considerations on the biomedical research area as a result of perceived vast commercial potential of genetic technologies. Primarily it highlights what role universities and their faculties will have in the commercialization process.

Specific questions addressed include:

First, what impact will involvement in commercialization have on the university research climate?

Second, with the remarkable numerical potential of commercial genetic products, who is entitled to financial return?

Third, if we establish that there might be problems with university involvement, what mechanisms are available to protect the interests of the university?

Fourth, how does biomedical research compare to other disciplines, where industry–university relationships have existed for some time?

Fifth, is it important to examine commercialization in light of existing mechanisms for commercialization?

Sixth, it is important that the issue of guidelines be addressed. Is it necessary for universities to develop codes of ethics for faculty members?

3) 16–17 June 1982—Hearing before the Subcommittee on Investigations and Oversights of the Committee on Science and Technology (U.S. House. 97th Congress, 2d session).

University/Industry Cooperation in Biotechnology.[65]

This forms part of the "Gore Committee's" ongoing investigation into perceived potential problems resulting from universities exploring research and commercial potential in the field of biotechnology, as a result of massive amounts of funding from industrial interests. The issues identified to be addressed by this hearing include:

First, what can be learned from current industry–university agreements in biotechnology?

Second, how are universities coping or planning to cope with difficult institutional and ethical questions?

Third, what are industries' needs in biotechnology and how do ties with universities satisfy these needs?

Fourth, what role should the Federal Government play in maintaining a balance between the need for technological innovation and the need to safeguard the university as a basic science research center?

4) 9 March 1982—Hearing before a Subcommittee on Energy, Nuclear Proliferation, and Government Processes of the Committee on Governmental Affairs (U.S. Senate. 97th Congress, 2d session).

Research and Development in the United States: The Role of the Public and Private Sector.[66]

This general investigative hearing recognizes the role of research and development in fostering a strong economic recovery. The thesis of Senator Charles H. Percy centers around the customed need for the federal government to support the national interest when private industry cannot justify the expenditures. Of particular relevance is the support of university research programs.

5) 2, 23 June 1982—Hearings before the Subcommittee on Investigations and Oversight of the Committee on Science and Technology (U.S. House. 97th Cong., 2d sess.).

Robotics.[67]

A portion of testimony centers on MIT's role in the development of robotics and the university's relationship with industry.

6) 1 March 1983—Introduction of S.631 by Senator Paul E. Tsongas, Democrat of Massachusetts. This bill entitled "High Technology Morrill Act" provides for the establishment of a national technology education grants program to provide matching Federal assistance to joint initiatives by private industry, educational institutions, and state government to strengthen science, engineering, and technical education.[68]

7) 1 March 1983—Hearings before Task Force on Energy and Technology and Task Force on Education and Employment of the Committee on the Budget (U.S. House. 98th Cong., 1st sess.).

Higher Education and Innovation in the U.S. Economy and the President's Fiscal Year 1984 Budget: Perspective from the States.[69]

Robert S. Rosenzweig, President of the Association of American Universities; Robert L. Sproull, President of the University of Rochester; and Frank H. T.

Rhodes, President of Cornell University, testify regarding the impact of federal budget reductions on scientific research and development at their universities and the potential role of joint industry–university interactions to assist in filling the gap.

Both the executive branch and the legislative branch of government view the industry–university research connection as a growing and significant aspect of our national economy. Administrations have differed in their philosophy of appropriate role for Government and the various means available to encourage and support these relationships. Some Congressional committees have questioned the appropriateness of these research ties with private industry and the ethical issues which might arise when large amounts of private funding are offered. Others see this trend as an important aspect of continued funding of our national research and development efforts.[70] It appears clear that these research initiatives originating between two parties at the exclusion of the federal government will continue for some time into the future. The question of the future role of the Federal Government in whether it attempts to regulate these types of agreements via legislation is still one of conjecture. Until an active Federal involvement through statute or administrative agency action comes into being, the use of contracting to establish and clarify these research interactions will continue to play a significant part of this research process. The law of industry–university research agreements, therefore, is currently based on existing common law, state and federal statutory law, and applicable administrative rules and regulations.

NOTES

1. W. H. Cowley, "Taxonomy of American Higher Education," W. H. Cowley Papers, Stanford University Library, Stanford, California.

2. A. Bartlett Giamatti, *The University and the Public Interest* (New York: Atheneum, 1981), p. 119.

3. U.S. Congress, House, Committee on Ways and Means, *General Revenue Revision, Hearing before the Committee on Ways and Means on Topics Pertaining to the General Revenue Revision of the Internal Revenue Code, Part 2,* 85th Cong., 2d sess., 1958, p. 1751.

4. Clark Kerr, *The Uses of the University*, 3d ed. (Cambridge, Mass.: Harvard University Press, 1982), pp. 46–48.

5. *Act donating Public Lands to the Several States and Territories which may provide Colleges for the benefit of Agriculture and Mechanical Arts*, ch. CXXX, 12 Stat. 503 (1862).

6. Earle D. Ross, *Democracy's College: The Land-Grant Movement in the Formative Style* (Ames, Iowa: Iowa State College Press, 1942), p. 47.

7. Ibid., p. 68.

8. Edward D. Eddy, Jr., *Colleges for Our Land and Time: The Land-Grant Idea in American Education* (New York: Harper & Brothers, 1957), p. 276.

9. See, generally, George N. Rainsford, *Congress and Higher Education in the Nineteenth Century* (Knoxville: University of Tennessee Press, 1972).

10. A. C. True, *A History of Agriculture Experimentation and Research in the United States, 1607–1925* [Misc. Pub. No. 251, U.S. Dept. of Agriculture] (Washington, D.C.: U.S. Government Printing Office, 1937).

11. See, generally, A. Hunter Dupree, *Science in the Federal Government: A History of Policies and Activities to 1940* (Cambridge, Mass.: Harvard University Press, 1957); Charles V. Kidd, *American Universities and Federal Research* (Cambridge, Mass.: Harvard University Press, 1959); and Dael Wolfle, *The Home of Science: The Role of the University* (New York: McGraw-Hill, 1972).

12. *Army Appropriations Act*, ch. 418 §2, 39 Stat. 649 (1916).

13. Bernard M. Baruch, *American Industry in the War: A Report of the War Industries Board* (Washington, D.C.: U.S. Government Printing Office, 1921); U.S. Food Administration, *Annual Report for the Year 1918* (Washington, D.C.: U.S. Government Printing Office, 1919).

14. Lewis E. Auerbach, "Scientists in the New Deal: A Pre-War Episode with Relations Between Science and Government in the U.S.," *Minerva* 3 (Winter 1965): 457–482.

15. Bruce L. R. Smith and Joseph J. Karlesky, *The State of Academic Science: The Universities in the Nation's Research Effort* (New York: Change Magazine Press, 1977), pp. 12–13.

16. Exec. Order No. 2859, May 11, 1918.

17. National Academy of Sciences, *Annual Report for 1918* (Washington, D.C.: U.S. Government Printing Office, 1919), pp. 40–41.

18. Lance E. Davis and Daniel J. Kevles, "The National Research Fund: A Case Study in the Industrial Support of Academic Science," *Minerva* 12 (April 1974): 207–220.

19. A. Hunter Dupree, *Science in the Federal Government*, p. 368.

20. Order by the Council of National Defense Establishing the National Research Committee, June 27, 1940, 5 Fed. Reg. 2446 (1940); Exec. Order No. 8807, dated June 28, 1941, 3 C.F.R. 959 (1938–1943 Compl), (as amended by Exec. Order No. 9309, dated October 18, 1943, 3 C.F.R. 1256 (1939–1943 Compl). See, generally, Irvin Steward, *Organizing Scientific Research for War: The Administrative History of the Office of Scientific Research and Development* (Boston: Little, Brown and Co., 1948).

21. See, generally, James Baxter, III, *Scientists Against Time* (Boston: Little, Brown and Co., 1946), which provides the official history of OSRD. Appendix C in that text lists twenty-five non-industrial contractors with OSRD and the twenty-five principal industrial contractors with OSRD.

22. Smith and Karlesky, *The State of Academic Science,* p. 13. See also, U.S. Senate, Committee on Military Affairs, Subcommittee on War Mobilization, *The Government's Wartime Research and Development 1940–44* [Subcommittee Report No. 5] (Washington, D.C.: U.S. Government Printing Office, 1945).

23. Vannevar Bush, *Science—The Endless Frontier: A Report to the President on a Program for Postwar Scientific Research*, July 1945 (Washington, D.C.: U.S. Government Printing Office: 1945), pp. 3–4.

24. U.S. Senate, Committee on Military Affairs, Subcommittee on War Mo-

bilization, *Report on Science Legislation: National Science Foundation* [Subcommittee Report No. 8] (Washington, D.C.: U.S. Government Printing Office, 1946).

25. Don K. Price, *Government and Science: Their Dynamic Relation in American Democracy* (New York: New York University Press, 1954), p. 50.

26. John R. Steelman, *Science and Public Policy: A Program for the Nation,* vol. I (Washington, D.C.: U.S. Government Printing Office, 1947), p. 6.

27. Smith and Karlesky, *The State of Academic Science*, p. 16.

28. See, generally, Walter S. Baer, "The Changing Relationships: Universities and Other R & D Performers," in *The State of Academic Science: Background Papers*, ed. B. L. R. Smith and J. J. Karlesky (New Rochelle, N.Y.: Change Magazine Press, 1978), pp. 63–72. See also, Herbert F. York and G. Allen Greb, "Military Research and Development: A Postwar History," in *Science, Technology and National Policy*, ed. T. J. Kuehn and A. L. Porter (Ithaca, N.Y.: Cornell University Press, 1981), pp. 190–215; W. Henry Lambright, *Governing Science and Technology* (New York: Oxford University Press, 1976), pp. 15–26; U.S. Congress, House, Committee on Government Operations, *Research and Development (Part 1—Background Testimony; Part 2—Military Research Representatives), Hearings before a Subcommittee of the House Committee on Government Operations*, 85th Cong., 2d sess., 1958.

29. Daniel S. Greenberg, *The Politics of Pure Science* (New York: New American Library, 1967), p. 273. See, generally, Donald W. Cox, *America's New Policy Makers: The Scientists' Rise to Power* (Philadelphia: Chilton Books, 1964), pp. 49–74.

30. 72 Stat. 1580 (1958).

31. Greenberg, *The Politics of Pure Science*, p. 274.

32. Barbara B. Clowse, *Brainpower for the Cold War: The Sputnik Crisis and National Defense Education Act of 1958* (Westport, Conn.: Greenwood Press, 1981), pp. 158–160.

33. James E. Katz, *Presidential Politics and Science Policy* (New York: Praeger Publishers, 1978). See also, Irving L. Horowitz and James E. Katz, *Social Science and Public Policy in the United States* (New York: Praeger Publishers, 1975).

34. Ibid., p. 260.

35. Claude E. Barfield, *Science Policy from Ford to Reagan: Change and Continuity* (Washington, D.C.: American Enterprise Institute for Public Policy Research, 1982), pp. 113–114. See also, Jon D. Miller, *The American People and Science Policy: The Role of Public Attitudes in the Policy Process* (New York: Pergamon Press, 1983).

36. See Edwin Mansfield, *The Economics of Technological Change* (New York: W. W. Norton & Co., Inc., 1968); Edwin Mansfield, *Industrial Research and Technological Innovation: An Econometric Analysis* (New York: W. W. Norton & Co., Inc., 1968); Edwin Mansfield et al., *Research and Innovation in the Modern Corporation* (New York: W. W. Norton & Co., Inc., 1971); Edwin Mansfield et al., *The Production and Application of New Industrial Technology* (New York: W. W. Norton & Co., Inc., 1977); Edwin Mansfield et al., *Technology Transfer, Productivity, and Economic Policy* (New York: W. W. Norton & Co., Inc., 1982).

37. See, generally, National Science Foundation, *Research and Development in Industry 1973* [NSF 75–315] (Washington, D.C.: U.S. Government Printing Office, 1974); National Science Foundation, *1985 R&D Funding Projections* [NSF 76–314]

(Washington, D.C.: U.S. Government Printing Office, 1976); National Science Foundation, *National Patterns of R&D Resources* [NSF 77–310] (Washington, D.C.: U.S. Government Printing Office, 1977); National Science Foundation, *Federal Funds for Research and Development* [NSF 79–310] (Washington, D.C.: U.S. Government Printing Office, 1979); National Science Foundation, *Federal Support to Universities, Colleges, and Selected Non-Profit Institutions: A Report to the President and Congress* [NSF 81–308] (Washington, D.C.: U.S. Government Printing Office, 1981); National Science Foundation, *National Patterns of Science and Technology Resources 1982* [NSF 82–319] (Washington, D.C.: U.S. Government Printing Office, 1982).

38. See, generally, Smith and Karlesky, *State of Academic Science*, pp. 16–19.

39. See National Science Foundation, *National Patterns of R&D Resources: Funds and Manpower in the United States 1953–1977* [NSF 77–310], (Washington, D.C.: U.S. Government Printing Office, 1977), p. 1.

40. Ibid., p. 4.

41. National Science Board, *University–Industry Research Relationships: Myths, Realities, and Potentials, Fourteenth Annual Report* (Washington, D.C.: U.S. Government Printing Office, 1982), pp. 5–7.

42. Jack Magarrell, "Corporate Chiefs Favor Giving to Higher Education, Survey Finds," *Chronicle of Higher Education* 24 (26 April 1982):7. See also, Kathleen Teltsch, "Gifts to Universities Rise 6.2%, Survey Finds," *New York Times*, 6 May 1984, sec. Y, p. 43.

43. National Science Foundation, *Academic Science: R & D Funds, FY 1980* [NSF 82–300]n (Washington, D.C.: U.S. Government Printing Office, 1982) p. 7.

44. David F. Noble, *America by Design: Science, Technology, and the Rise of Corporate Capitalism* (New York: Alfred A. Knopf, 1977), p. 147.

45. Gilbert S. Omenn, "Re-Energizing the Research University," *Environment* 24 (July/August 1982):50.

46. Ibid.

47. See, generally, the three part series "The Academic-Industrial Complex": "Can Industry Control Campus?" *St. Louis Post-Dispatch*, 26 September 1982, p. A–13; "Swift Negative Reaction Shelves Harvard's Plan," *St. Louis Post-Dispatch,* 27 September 1982, p. 1; "Biomedical Research 'Marriage' Here," *St. Louis Post-Dispatch*, 28 September 1982, p. A–1.

48. Denis J. Praeger and Gilbert S. Omenn, "Research, Innovation, and University–Industry Linkages," *Science* 207 (25 January 1980):380.

49. Ibid., p. 381.

50. National Commission on Research, *Industry and the Universities: Developing Cooperative Research Relationships in the National Interest* (Pasadena, Calif.: National Commission on Research, August 1980), pp. 20–23. See, generally, Donald R. Fowler, "A Study of the Need for and the Impediments to Improved and Novel University–Industry Research Relationships" (Ph.D. dissertation, Claremont Graduate School, 1983), p. 4; Praeger & Omenn, "Research, Innovation, and University–Industry Linkages," p. 381. B. L. R. Smith and J. J. Karlesky, *The State of Academic Science* (1977), pp. 65–66; Rustum Roy, "University–Industry Interaction Patterns," *Science* 178 (1 December 1972):955–960.

51. Edward E. David, Jr., "Supporting Research with a Commercial Mission,"

Change 14 (September 1982):26–29. See also, Roy V. Gavet, "Business and Academia: An Emerging Partnership," *Change* 15 (April 1983):23–28.

52. George Keller, *Academic Strategy: The Management Revolution in American Higher Education* (Baltimore: Johns Hopkins University Press, 1983), pp. 19–22.

53. See, generally, James Botkin, Dan Dimancescu, and Ray Stata, *Global Stakes: The Future of High Technology in America* (Cambridge, Mass.: Ballingen Publishing Co., 1982), pp. 113–137.

54. President's Message to Congress on Science and Technology, 8 Weekly Comp. Pres. Doc. 587 (16 March 1972).

55. National Science Foundation, *Thirty-First Annual Report for Fiscal Year 1981* (Washington, D.C.: U.S. Government Printing Office, 1982), pp. 75–78. See also, Sumner Myers and Eldon E. Sweezy, *Federal Incentives for Innovation: Why Innovations Falter and Fail; A Study of 200 Cases,* Report R–75–04 (Washington, D.C.: National Science Foundation, 1976).

56. J. N. Nay and J. Waller, *Regulatory Administrative Experiment Manual* (NTIS PB 298446, June 1979); "Industry Incentives Program Underway," *Dimensions NBS* 58 (February 1974):29.

57. 94 Stat. 2311 (1980).

58. U.S., Congress, House, Committee on Science and Technology, *Government and Innovation: University–Industry Relations, Hearings before a Subcommittee on Science, Research, and Technology*, 96th Cong., 1st sess., July 31, August 1, 2, 1979.

59. 94 Stat. 3015 (1980). See, generally, U.S. Congress, House, Committee on the Judiciary, *Industrial Innovation and Patent and Copyright Law Amendments, Hearing Before a Subcommittee on Courts, Civil Liberties, and the Administration of Justice*, 96th Cong., 2d sess., April 3–June 9, 1980.U.S. Congress, Joint Hearing before the Senate Committee on Commerce, Science, and Transportation and the House Committee on Science and Technology, *Federal Patent Policy*, 97th Cong., 1st sess., September 30, 1981.

60. U.S. Department of Justice, Antitrust Division, *Antitrust Guide Concerning Research Joint Ventures* (Washington, D.C.: Government Printing Office, 1980).

61. 95 Stat. 172 (1981).

62. U.S. Congress, House, Committee on Science and Technology, *The Government-University Accountability Relationships in the Field of Scientific Research, Hearing before the Subcommittee on Science, Research, and Technology*, 96th Cong., 2d sess., March 24, 1980.

63. 92 Stat. 3 (1978).

64. U.S. Congress, House, Committee on Science and Technology, *Commercialization of Academic Biomedical Research, Hearings before a Subcommittee on Science, Research, and Technology*, 97th Cong., 1st sess., June 8, 9, 1981.

65. U.S. Congress, House, Committee on Science and Technology, *University/Industry Cooperation in Biotechnology, Hearing before the Subcommittee on Investigations and Oversights*, 97th Cong., 2d sess., June 16, 17, 1982.

66. U.S. Congress, Senate, Committee on Governmental Affairs, *Research and Development in the United States: The Role of the Public and Private Sectors, Hearing before a Subcommittee on Energy, Nuclear Proliferation, and Government Processes*, 97th Cong., 2d. sess., March 9, 1982.

67. U.S., Congress, House, Committee on Science and Technology, *Robotics,*

Hearings before the Subcommittee on Investigations and Oversight, 97th Cong., 2d sess., June 2, 23, 1982.

68. U.S. Congress, Senate, Committee on Labor and Human Resources, *Education for Economic Security, Hearing before a Subcommittee on Education, Arts, and Humanities*, 98th Cong., 1st sess., March 8, 9, April 8, 1983.

69. U.S., Congress, House, Committee on the Budget, *Higher Education and Innovation in the U.S. Economy and President's Fiscal Year 1984 Budget: Perspective from the States, Hearings before Task Force on Energy and Technology and Task Force on Education and Employment*, 98th Cong., 1st sess., March 1, 1983.

70. See U.S. Congress, Office of Technology Assessment, *Commercial Biotechnology: An International Analysis* (Washington, D.C.: U.S. Government Printing Office, OTA-BA–218, 1984).

3

Major Legal Issues Relating to Sponsored Research Agreements

TRADITIONAL FUNDAMENTAL RESEARCH AND THE INTERESTS OF INDUSTRY

The research university performs at its best in the area of fundamental research. This traditional role while remaining aloof from the market place remains crucially significant to the future of commerce and industry as a whole. Although this description is somewhat oversimplified, individual corporations tend to concentrate in two categories of interactions with academic research institutions. These have been identified by Robert M. Rosenzweig, Vice President at Stanford University, as follows:

> The first is broadly philanthropic. Its extent has grown substantially in recent decades, and it expresses the broadest interest of the business community in the health of colleges and universities. The second is more focused, in that it rests as a view of institutions of higher learning as the producers of products of value.
>
> The most important of these are educated persons and research of more or less current utility. . . . The most productive research is likely to arise from relationships forged among scientists in one company and in one university.[1]

In March 1982 the heads of five major research universities and eleven corporations met in seclusion at Pajaro Dunes on the coast of California to discuss academia's increasing interest in collaborative research efforts with industry. Particular emphasis was placed on biotechnology research.[2]

From this conference came a "draft" that the participants hoped would advance the debate on the topic of industry–university interactions while advancing the commercialization of biology. The theme of the conference emphasized collaboration between universities and industry as benefiting all parties, while not distorting the university's ideals with industry's vast financial resources. Use of research agreements was specifically cited in the draft as the primary means to achieve this goal:

> It is important that universities and industries maintain basic academic values in their research agreements. Agreements should be constructed, for example, in ways that do not promote a secrecy that will harm the progress of science, impair the education of students, interfere with the choice by faculty members of the

scientific questions or lines of inquiry they pursue, or divert the energies of faculty members from their primary obligations to teaching and research.[3]

The "draft" noted that the universities have the duty to see that academic values are not threatened by research relationships. To assure that they are maintained several steps were recommended: make public the provisions of research contracts; use faculty committees to review contracts to assure that the agreement is consistent with academic values; utilize other methods of review considered to be effective by the individual university.

In addition to contract disclosure or use of a committee to examine contracts, the report indicates that certain other legal issues should be highlighted. An active patent policy by the university was favored, even though filing may require a brief delay in publication or other public disclosure of research. No consensus was reached regarding the grant of an exclusive license to industry to develop a patent for profit. The Pajaro Dunes Statement tends to favor exclusive licenses in certain circumstances:

Although, in some cases, multiple licenses will undoubtedly speed development, in other cases, exclusive rights are essential if development is to take place, since no firm will expend large sums for development if others can reap the fruits.

Thus, universities should be able to negotiate exclusive licenses provided that exclusivity seems important to allow prompt vigorous development of the patent to occur. The desirability of exclusivity in certain cases is recognized under current federal law. When exclusivity is allowed, however, it should be permitted for only the interval as necessary to encourage the desired development. In addition, the university should insist upon a requirement of due diligence on the part of the licensee in developing and using the patent.[4]

Lastly, legal and ethical issues involving conflict of interest were addressed. The two aspects of the conflicts question centered on the propriety of a university taking an equity position in a company in which one of its faculty has become a major stockholder or officer. This was considered inadvisable for universities unless sufficient safeguards are instituted to avoid adverse effects on morale. No consensus was stated regarding a conflict of loyalties when faculty become affiliated with a biotechnology firm. Resolution of the conflict-of-interest issue was left for each individual university to handle according to its own circumstances and traditions. Although recognized as a broad based summit meeting, out of the Pajaro Dunes Conference came a general recognition of the significance of the various legal issues inherent in any industry–university research relationship.

IMPEDIMENTS TO RESEARCH RELATIONSHIPS

One of the foremost writers in the field of industry–university research relationships is Donald R. Fowler, General Counsel at the California Institute of Technology. In his 1983 survey of administrative attitudes toward research agreements between higher education and industry, Fowler listed those significant impediments which his research identified or which were verified as significant features in his literature review. Fowler's empirical survey identifies fifteen impediments of which only ten were regarded as significant by industry and only eight were regarded so by university respondents. He concludes that no overwhelmingly important impediments seem to control or dominate the university–industry research relationship. Fowler's study reports that the factor which presented the greatest concern to universities involved the need to protect the university's right to publish versus industry's need to protect proprietary information. He notes:

> [T]he university respondents overwhelmingly joined the industry respondents in declaring that universities should agree with industrial sponsors to withhold research results from publication during the time necessary for the university or industry to obtain patent protection.[5]

Reseach concerns such as this one which are successful as parts of an ongoing research project are stated and memorialized in the form of a legally binding agreement or contract.

The ideal research-and-development contract between an industry and a university must be mutually beneficial. The contracts must be carefully considered and negotiated to insure that the legitimate needs and objectives of all parties are met. In reference to all categories of impediments, research agreements must strike a balance reflecting the interest of both parties. Attorneys John W. Wilson, Jr., Richard P. Dobb, and William T. Gerl of the Georgia Institute of Technology address this goal:

> In this [negotiation] process, . . . companies have a vital interest in obtaining answers to three questions concerning the protection and ownership of technology developed under company/university R & D contracts:
>
> - Can the university protect the firm's proprietary and confidential information from disclosure?
> - Does the company own the rights to technology developed by the university in carrying out the R & D contract, or does the university?
> - Is the university willing to refrain from or defer publishing the results of its R & D if non-disclosure proves necessary to protect the company's competitive position?[6]

The following sections discuss the significant legal issues that contracts and agreements between industry and universities should address. This area of research will concentrate on both common law and statutory issues. Subsequent chapters will look at specific contracts and their formats to illustrate concerns that the contracting parties have identified and attempted to address in the language of the legal instrument.

CONTRACT LAW AND THE COMMON LAW OF HIGHER EDUCATION

Sponsored research agreements represent the most frequently accepted type of major research interaction between institutions of higher education and industry. In their commonly accepted form, they typically consist of an industry funded research activity involving a specific project, product, or other narrowly defined area. The agreements can involve basic research, applied research, or a combination of both. A variety of relationships can be created involving individual faculty, specific departments, university research groups, and the individual corporate entity.

When an institution of higher learning enters into a sponsored cooperative research agreement, the agreement typically manifests itself either as one in which the industrial sponsor may desire to enter into a partnership relationship to carry out a collaborative research effort or as one in which the university may seek to perform research geared to a specific good or product. Actual arrangements may blend these two formats. Specifically excluded from the scope of this analysis are lesser contracts, such as those between individual faculty and industry for consulting arrangements, industrial research funding given as a gift, and industrial associate agreements.

Fowler has identified fifteen categories of legal issues which recur generally in drafting research agreements:[7]

1. The Scope of the Research Project
2. Nature and Extent of the Sponsor's Commitment to the Project
3. Nature and Extent of the University's Undertaking Pursuant to the Agreement
4. Control Over the Conduct of the Funded Research Program
5. Exclusive Right of the Industrial Sponsor to Fund Research in the Area Involved in the Agreement
6. The Extent and Terms of Actual Technical or Scientific Collaboration by the Industry Participants
7. Reporting Requirements
8. Funding
9. Competing Interests in the Use of Research Results
10. Receipt of Proprietary Information from Industrial Sponsor
11. Patent Rights
12. The Licensing of "Know-How"
13. Indemnification and Hold Harmless Agreements
14. Use of University's Name
15. Potential Conflicts of Interest on the Part of the University Researchers

The discussion of legal issues that follows reflects Fowler's categories. Major legal questions only are presented with little or no discussion being given to the mechanics of a law's operation. In addition, an examination of areas of the law that are generally common knowledge are avoided. Since most of these contracting sections can be classified as falling either within case law, business law, or statutory law, they have been organized to reflect this division.

The Scope of the Research Project

Scientific and engineering research, whether basic or applied, is often of a speculative nature. When negotiating their research arrangement the university scientist and a potential industrial sponsor are frequently unclear about the specific scope and bounds of the project, as well as details about a desired result. Scientific results are traditionally unpredictable. This lack of predictability inherent in scientific research requires that any research arrangement be negotiated with a maximum of specificity and precision. Fowler points out that many aspects of the arrangement other than a desired output of the research depend on accurate descriptions. He states the following:

> For example, control over the research project, any right of the industry sponsor to exclusive funding of the specific area of research, and patent or publication rights often turn on the defined scope of the research project. Failure to have a well-defined scope can lead . . . to embarrassing and costly disputes over who owns the patent rights to inventions made by the researcher . . . during the term of the research agreement, but not necessarily as part of it.[8]

The Sponsor's Funding Commitment to the Project

Year-to-year funding is a common option with which most institutions of higher education are familiar. The majority of universities operate on a fiscal year basis and also have experience with federal government sponsored research funded annually. If year-to-year funding is a compatible option, it should be considered as a contracting provision.

More often, however, an industrial sponsor would prefer to fund a research project on a continuing basis without the need for formal annual renewals, especially if the research is progressing and continues to fit the firm's research and marketing goals. Similarly, the university may require greater funding stability and predictability. After making a commitment to new staff, physical plant, and equipment, which may not otherwise be required, the corporate sponsor can expect that the university will require assurances that the sponsor will not abandon its interest in a successful ongoing project to continue work on its own. One method for assuring progress is to provide in the contract that the university must meet the industry performance standards prior to further financial support. Fowler suggests negotiating a provision by which the industrial sponsor agrees to fund the overall project within limitations as long as the project meets clearly defined checkpoints:

> The ability to establish and agree upon such intermediate goals or "gates" depends on the degree of specificity to which the research project, and its intermediate and ultimate goals, lend themselves. Although often difficult to set forth, intermediate "gates" should be identified in the agreement with a provision whereby the sponsor agrees to continue funding the project, at agreed upon levels, as long as the research continues to meet those targets, until either the time or funding negotiated for the total project have [*sic*] expired.[9]

Nature and Extent of the University's Commitment

Since researchers recognize that they cannot predict with precise certainty the outcome of their scientific endeavors, a university should not enter into a contractual relationship with industry that will guarantee a predetermined result. Guaranteeing results conflicts with the inherently flexible nature of scientific research and violates principles of conducting that research. Any detailed expectation would be unethical and would tend to exploit the university's good name. Specific clauses requiring firm commitments in university scientific research must be avoided.[10]

Universities are nonprofit, tax-exempt, charitable corporations. By their very charter of incorporation, they are considered institutions that seek to avoid financial risk. This classification dictates that the university, unlike its industrial counterpart, has no legal basis on which to place its funds at risk by financing speculative proprietary research arrangements. It may receive private funding for these purposes but may not commit its own resources to such ventures.

Liability for unauthorized or improper expenditures could be severe. The university could be liable for improper diversion of charitable funds. As pointed out in George G. and George T. Bogerts' treatise on trusts, in many states recent revisions of statutory law of the general standard of care applied to a typical "trustee" of a university organized as a nonprofit corporation tend to closely match the standard imposed on the director of a general business corporation.

> The power, duties, and liabilities of trustees for charity are, with only a few exceptions, the same as in the case of private trusts. . . .
>
> Just as in the case of a private trust, so in the case of a charity the trustee owes a duty of loyalty. . . .
>
> If a trustee for charity improperly spends trust monies, he may be obliged to restore such sums in a suit brought by the Attorney General.[11]

Discussions in legal literature as to the differences in the duties of charitable trustees and directors or trustees of a charitable corporation show some uncertainty in the law. A report sponsored by the Ford Foundation on the question of delegation of investment responsibility concluded that a trustee is not permitted to delegate to others the making of investment decisions for the trust. Directors of a charitable corporation, however, may delegate investment responsibilities to committees or responsible individuals within the institution or to outside investment counselors, but proper supervision must be exercised.[12]

The federal courts have examined this issue and identified the underlying problems:

> The applicable law is unsettled. The charitable corporation is a relatively new legal entity which does not fit neatly into the established common law categories of corporation and trust. As the discussion below indicates, however, the modern trend is to apply corporate rather than trust principles in determining the liability of the directors of charitable corporations, because their functions are virtually indistinguishable from those of their "pure" corporate counterparts.[13]

The Bogerts' text indicates that trustees are held to a higher standard of care and are liable for simple negligence, whereas a director is required only to exercise ordinary and reasonable care in the performance of his duties and will be liable only if "gross negligence" has been committed.[14] Thus, the strict liability standard imposed on the trustee of a true trust does not generally apply to these non-profit institutions.[15] For example, under California's Nonprofit Corporation Law the not-for-profit trustees for the purpose of investment management are held to a "prudent man" standard of care.[16]

Unauthorized and improper expenditures for proprietary research may result in the loss of an institution's tax exempt status if the institution allows the results of its operations to inure to the benefit of a private party. According to Section 501 (c) (3) of the *Internal Revenue Code* (I.R.C.), exemption from federal income tax allowed by Section 501 (a) will apply only to those organizations "no part of the net earnings of which inures to the benefit of any private shareholder or individual."[17] Chief Justice Warren Burger recently restated this proposition:

> Charitable exemptions are justified on the basis that the exempt entity confers a public benefit—a benefit which the society or the community may not itself choose or be able to provide, or which supplements and advances the work of public institutions already supported by tax revenues. History buttresses logic to make clear that, to warrant exemption under § 501 (c) (3), an institution must fall within a category specified in that section and must demonstrably serve and be in harmony with the public interest.[18]

Unlike many contract provisions between for-profit corporate entities, the first provision to be considered for an industry–university research agreement should state that the research will be carried out by the university on a "best effort" basis with no financial or other contractual penalty for default. The courts have explained the "best effort" concept thus:

"Best effort" imposes a legal duty of performance more demanding than mere competence or due diligence and . . . it means maximizing the contractual benefits of the person to whom the duty is owed. . . . It involves a stricter standard of performance than good faith.[19]

The corporate sponsor's remedy rests in its contractual right to withdraw its support from the project. Those corporations negotiating with universities must do so from the stance that typical firm performance commitments, fixed prices, and contractual provisions to recoup damages in case of default are foreign and inappropriate to the non-profit and tax exempt educational institution.

Faculty and Institution Academic Freedom and Corporate Control Over Research Programs

Industrial sponsors who have experience in funding university research efforts are familiar with the university's demand for unrestricted project selection and non-interference in its approach to scientific inquiry. With the exception of requirements for technical reporting and collaboration, the concept of "academic freedom" often appears to the industrial sponsor as relinquishing all control to the university and its researchers. This appearance of "heads, I win; tails, you lose" should not necessarily create a divergence from many corporate research efforts where internally employed scientists exercise a good deal of independence.

This tradition of universities' avoiding any appearance of control by the industrial sponsor has its roots in both the historical common law and university tradition, as well as a constitutional law nexus. No legal definition of academic freedom exists, and there is no uniform agreement among scholars as to its precise meaning.[20] A frequently quoted definition is one made by Arthur O. Lovejoy, one of the founders of the American Association of University Professors (A.A.U.P.):

Academic freedom is the freedom of a teacher or researcher in higher institutions of learning to investigate and discuss the problems of his science and to express his conclusions, whether through publication or the instruction of students, without interference from political or ecclesiastical authority or from administrative officials of the institution in which he is employed, unless his methods are found by qualified bodies of his own profession to be clearly incompetent or contrary to professional ethics.[21]

Ralph F. Fuchs, a subsequent president of the A.A.U.P., stated that academic freedom has keystones to its foundation:

1. the philosophy of intellectual freedom, which originated in Greece, arose again in Europe, especially under the impact of the Renaissance, and came to maturity in the Age of Reason;
2. the idea of autonomy for communities of scholars, which arose in the universities of Europe; and
3. the freedoms guaranteed by the Bill of Rights of the Federal constitution as elaborated by the courts.[22]

The legal concept of academic freedom has two branches: first, the personal freedom of the individual and second, the institutional autonomy of the university.[23] While the former concept is well established, even though its contours may be uncertain, the latter continues to struggle for recognition.[24]

In the private university, the employment contract manifests itself as the principal source of the faculty member's legal rights.[25] Usually, the contract for faculty employment contains no mention of academic freedom. But it tends to be understood, at least in the major universities, that the terms of employment incorporate the policy statements that appear in faculty handbooks, often stated in terms adopted by the A.A.U.P., as well as the general mores and traditions of the academic community.[26] The A.A.U.P. 1940 statement, as amended, on academic freedom pertinent to the research interests involved in industry–university agreements states:

Academic freedom

(a) The teacher is entitled to full freedom in research and in the publication of the results, subject to the adequate performance of his other academic duties; but research for pecuniary return should be based upon an understanding with the authorities of the institution.

(b) The teacher is entitled to freedom in the classroom in discussing his subject, but he should be careful not to introduce into his teaching controversial matter which has no relation to his subject. Limitations of academic freedom because of religious or other aims of the institution should be clearly stated in writing at the time of the appointment.

(c) The college or university teacher is a citizen, a member of a learned profession, and an officer of an educational institution. When he speaks or writes as a citizen, he should be free from institutional censorship or discipline, but his special position in the community imposes special obligations. As a man of learning and an educational officer, he should remember that the public may judge his profession and his institution by his utterances. Hence he should at all times be accurate, should exercise appropriate restraint, should make every effort to indicate that he is not an institutional spokesman.[27]

In the public university, the individual faculty member's right of academic freedom finds its protection in the U.S. Constitution, as well as in the employment contract. The Constitutional basis of protection originates from either the First or Fourteenth Amendment, depending on the alleged grievance. The Constitutional shield is available to faculty employees of the public university more so than to those of the private university, as the Constitutional Amendments apply to state action.[28] A recent opinion by the Seventh Circuit, *Dow Chemical v. Allen*,[29] holds that the same First Amendment considerations may be a valid basis for withholding information about scientific research.

In *Dow Chemical*, the manufacturer had subpoenaed the working papers, notes, and data of University of Wisconsin research investigators who were studying the side-effects of the herbicide Agent Orange produced by Dow. The question presented was

> whether a private corporation, Dow Chemical Company, threatened with possible government cancellation of certain herbicides it manufactures, may compel through administrative subpoenas University of Wisconsin researchers to disclose all of the notes, reports, working papers, and raw data relating to on-going, incomplete animal toxicity studies so that it may evaluate that information with a view toward possible use at the cancellation hearings.[30]

In holding that the facts of the case did not warrant forced disclosure of university research information, the court noted, after citing *Bakke*,[31] *Sweezy*,[32] *Cooper*,[33] *Barenblatt*,[34] and *Keyishian*,[35] that

> enforcement of the subpoenas would leave the researchers with the knowledge throughout continuation of their studies that the fruits of their labors had been appropriated by and were being scrutinized by a not-unbiased third party whose interests were arguably antithetical to theirs. It is not difficult to imagine that that realization might well be both unnerving and discouraging. Indeed, it is probably fair to say that the character and extent of intervention would be such that, regardless of its purpose, it would "inevitably tend to check the ardor and fearlessness of scholars, qualities at once so fragile and so indispensable for fruitful academic labor." In addition, the researchers could reasonably fear that additional demands for disclosure would be made in the future. If a private corporation can subpoena the entire work product of months of study, what is to say further down the line the company will not seek other subpoenas to determine how the research is coming along? To these factors must be added the knowledge of the researchers that even inadvertent disclosure of the subpoenaed data could jeopardize both the studies and their careers. Clearly, enforcement of the subpoenas carries the potential for chilling the exercise of First Amendment rights.
>
> . . . [O]ur point is simply that respondents' interest in academic freedom may properly figure into the legal calculation of whether forced disclosure would be reasonable. . . . We conclude there is little to justify an intrusion into university life which would risk substantially chilling the exercise of academic freedom.[36]

Dow Chemical has interesting overtones. While it aligns academic freedom with scholarly output, where some authorities believe the concept properly belongs, it also associates academic freedom with control over the research product, which arguably is a departure from previous concerns. Historically, the concept of academic freedom has dealt with the right to express controversial opinions, free from the discomfort or dangers of censorship, harassment, ostracism, or even death. The *Dow Chemical* court, though, appears more concerned with protecting careers from the insinuation of incompetence, a point not overlooked by Judge Pell, who concurred in the result but dissented from that part of the opinion designated as "academic freedom":

The respondents are concerned, which concern is also expressed in the opinion of this court, that periodic disclosure of the data could severely jeopardize the careers and reputations of the researchers. I do not share in this concern as at the time the subpoenas were issued, as I understand it, plaintiffs were just seeking laboratory data. I can see that researchers might draw early conclusions from data which was not substantiated by further empirical laboratory studies. The public exposure to the early conclusions of the researchers might well cast some doubt on their ability to analyze laboratory data. What is involved here, however, seems to me to be merely a matter of recording accurately. A researcher's reputation perhaps deserves to be subject to some questioning if he or she cannot accurately observe and record specific factual matters.[37]

Judge Pell's opinion touches on another issue of potential importance in research contracts regarding who actually has the right to control release of the work product. In this case, the judge questions whether this control should rest with the investigators or the agency funding the research (in this case, the Department of Health, Education, and Welfare):

I do not intend in any way to belittle the importance of academic freedom in the society of this nation. I am also mindful, however, that this was not an independent investigation engaged in by faculty researchers and financed by the University. The research is being conducted pursuant to a grant application to the Department of Health, Education, and Welfare. The fact that it is being financed by government money, I assume, would not mean that the ongoing study was automatically subject to public disclosure at each step and stage thereof. Nevertheless, it is quite reasonable to assume that the undertaking of the study was directly limited to whether or not the product being researched should continue in the public marketplace. The chemical involved and its possible deleterious effects on general health, and particularly on reproductive efficiency, had received widespread public attention following the Viet Nam War. The grant application for the initial stage of the study had a proposed beginning date of July 1, 1977.The cancellation hearing notice, which ultimately resulted in the subpoenas under review, was issued on February 1979. It seems clear to me that before this court purports to pass on the academic freedom issue, . . . it should be supported by a more complete factual record and a more complete disclosure and discussion by the parties involved than it now has.[38]

The academic community has traditionally been sensitive to the threat of interference or control from outside sources. As reflected in contemporary historical issues, this apprehension has crystallized in three areas:

First, the post World War II legislative investigations of the McCarthy era with their loyalty oaths, led to heightened concern for the protection of free expression;

Second, the injection of large sums of ear-marked federal dollars led to the worry that the dignity and independence of the university would be sacrificed in competing for those funds;

Third, the intervention of the federal government in hiring decisions and policies of the past decade has spurred lively debate over whether the modern university has a claim to institutional autonomy.[39]

Control over funding and the academic institution's concern and distrust of outside funding is not a new phenomenon. During the industrial expansion in the post Civil War reunion years of the latter part of the 19th century, philanthropists became conspicuous contributors to higher education. This development was marked by both cheer and gloom by the university community. This phenomenon was noted by historians Richard Hofstadter and Walter Metzgar as follows:

The patrons of the university received from the academic world the ornate courtesies of gratitude. They did not enter academe as intruders; they were welcomed into the realm and escorted to its high places by its very grateful inhabitants. Within the academic fraternity, to cultivate the good will of donors was a highly approved activity, betokening fine public spirit. To offend the bearer of gifts was an action sometimes defined as the deepest disloyalty and treachery. Cordiality was thus demanded of professors by the most compelling of motives—self-interest and the desire for social approval.

In the light of these reasons for friendship, it is particularly surprising that sharp antagonisms developed over the issue of academic freedom. Yet almost from the moment of confrontation, the picture of the business patron as an enemy of academic freedom took form in the minds of professors. This began in the middle eighties, when Professor Henry Carter Adams was dismissed from Cornell for having delivered a pro-labor speech that annoyed a powerful benefactor.[40]

This continuing institutional suspicion, which traditionally has been directed primarily at the federal government and the private foundations, has become more muted but perhaps no less keen as industry–university research agreements increase in number. Russell Kirk states the concern as follows:

For no proverb is truer than this, that "The man who pays the piper calls the tune." If educational institutions become dependent for their increase of reputation, or perhaps even for their existence, upon a few sources of benefaction, ineluctably most administrators and even professors will play their pipes accordingly. And this is not the less true merely because a government is "democratic," or a foundation "charitable." Governments and foundations are directed by men, like all of us, with prejudices and interests—which may not always be identical with the opinions and advantages of the more lively spirits in the Academy.[41]

Fuchs, in his role as A.A.U.P. President, states the issue thusly:

Inroads upon autonomy in respect to research are a leading cause of concern in American colleges and universities at present, because grants from government and industry for designated projects may influence the directions of inquiry. Here institutional integrity and individual self-direction both stand in need of protection—not from hostile action but from temptation.[42]

But not all opinions on the subject are negative. With regard to the rules attached to federal grants, one former high-level government administrator has said:

While universities have a clear obligation to observe the federal rules that exist, they have an equal right and obligation to protest existing or proposed terms and conditions which they find objectionable. Their effectiveness in controlling the terms and conditions under which they will accept federal research funds is the measure of their independence and freedom.[43]

The sponsors of industry–university research agreements appear to be attentive to this touchy issue of control. For instance, Howard Goodman, the noted biochemist who is responsible for initiating the $70 million Hoechst agreement with Massachusetts General Hospital, is said to be "adamant" on the subject of control over research. In an interview with *Science*, Goodman stated:

Hoechst has no influence on the direction of research. . . . Contractual legalese aside, as far as I'm concerned, this is a grant. This department is not an industrial extension.[44]

Institutional control over administrative decisions is essential and well established. The U.S. Court of Appeals for the Seventh Circuit recently observed there is a "policy of fostering academic freedom at the university level."[45] The concurring opinion of Judge Coffey elaborated more fully on the issue of institutional academic freedom and the need of institutions of higher education to control the essential elements that are fundamental and basic to university administration:

It is true that in recent years courts have increasingly intervened in school disciplinary situations involving major sanctions. The facts in this case, as the majority points out, can be considered as analogous to an academic rather than a disciplinary dismissal. In this area courts have always been, and continue to be, reluctant to intrude. It is my belief that the University of Wisconsin Law School and other academic institutions, *not* federal judges, are more qualified to make sensitive academic judgements as to their faculty's make-up as well as who should be admitted to study and upon what conditions they shall be admitted. If we were to impose the guiding hand of the federal judiciary into such decisions, we should diminish the vital precept of academic freedom to an oft-recited but empty cliche; one without meaning or substance. Basic academic decisions, such as the determination as to the make up of the faculty and who may be a student on the first day of classes, have long been regarded as among the essential prerogatives and freedoms of the university administration. Should we ever conclude otherwise, we would overstep our bounds into an area of academia in which we are ill-prepared to act and would ill-advisedly impinge upon the right of the administration to make fundamental and basic decisions as to the composition of their faculty and student body.[46]

Judge Coffey appears to suggest that universities merit a certain degree of institutional autonomy with regard to the academic decisions that affect both the acceptance and dismissal of students and the hiring and firing of faculty employees. Arguably this autonomy should extend also to control over research and research contracting.

Supportive arguments for the recognition of university autonomy, which might entail an evidentiary privilege, tend to rely on Justice Powell's discussion of the university in his plurality opinion in *Bakke*:

Academic freedom, though not a specifically enumerated constitutional right, long has been viewed as a special concern of the First Amendment. The freedom of a university to make its own judgments as to education includes the selection of its student body. Mr. Justice Frankfurter summarized the "four essential freedoms" that constitute academic freedom:

"It is the business of a university to provide that atmosphere which is most conducive to speculation, experiment and creation. It is an atmosphere in which there prevail 'the four essential freedoms' of a university—to determine for itself on academic grounds who may teach, what may be taught, how it shall be taught, and who may be admitted to study."[47]

The language is of special importance because the question of academic freedom is placed in the context of the university's "business," not the individual right of the faculty member, as in earlier cases.

The historical development of the concept of freedom of the university to pursue scientific inquiry without interference has its traditional foundation in the philosophy of academic freedom. Case law has traditionally supported and expanded these concepts. Individuals drafting a research agreement must be cognizant of these developments and not attempt to draft language into a contract which would violate this traditional research autonomy.

Protecting the University from Liability: Indemnification from Third Party Claims

As has been mentioned earlier, non-profit universities seek to avoid taking risks. When entering into contractual relationships with industry, universities must insist that liability risks be passed along to the profit motivated party of the contractual relationship. Use of indemnification concepts in the contract is the most typical method of seeking protection. Indemnification as a legal concept does not suggest an exemption from liability as do the doctrines of sovereign or charitable immunity. Instead, it provides a means for determining who should pay damages as between two or more likely parties.[48] The concept of indemnification may be based on principles of contract, equity, or agency. In contract law, indemnification refers to a bargained for agreement as to which party will assume the risk of liability.[49] In equity, indemnity refers to the right to restitution.[50] And in agency law, the doctrine imposes a promise of indemnity from the principal to the agent where damages arise out of a good faith execution of the agency relationship.[51]

Indemnification by Contract. An indemnity agreement is a form of insurance typically applicable to three party situations.[52] An indemnity contract is an agreement to make good and hold another harmless from financial loss on some obligation which he has incurred to a third party.[53] The right to indemnification may run to the victim or the party who caused the injury, and in the latter case it may be triggered by an alleged liability or an actual showing of loss. A special form of indemnification contract is the "hold harmless" agreement, whereby "one party assumes the liability of another . . . as an incidental obligation in contracts dealing with services, premises, or products."[54] An indemnification agreement, though, cannot cover liability arising from action that is contrary to public policy (e.g. civil rights violations).[55] The need for indemnification clauses in contracts between industry and universities is essential in today's medical research relationships. The growing number of lawsuits centered around the drug DES is an example of the type of time consuming and extended litigation returning to haunt the campus decades after the experiment has been concluded.[56] In research situations involving smaller corporations which might not exist when an injury results years later, the university should consider use of an insurance contract to cover potential liability.

Several states have adopted statutory indemnification provisions which protect public officials and employees from the risk of personal liability in an action arising out of employment duties.[57] Private employers may similarly protect their employees, as has been recommended in proposals for university risk management.[58]

The construction of indemnity contracts appears to be the source of considerable litigation, and several observations can be briefly stated. First, no unusual complications appear to characterize cases that involve the university as party, with the exception of the immunity defense which, if applicable, forecloses litigation for the university but not necessarily the employee.[59] Second, the meaning of contract language is a question of law.[60] Third, the rules pertaining to indemnity vary from state to state, so that conflict of laws may become a relevant issue in contract interpretation.[61] Fourth, although the indemnitee may contract away liability for his own negligence, unless prohibited by state statute or where a public duty is owed, the indemnitor's obligation must be expressed in clear and unequivocal language.[62] And finally, jurisdictions differ as to whether contract terms are to be construed strictly or broadly. Where strict construction is the rule, courts appear concerned with protecting the indemnitor from obligations he did not intend to undertake.[63] Contrarily, the purpose of liberal rules of construction is to protect the reasonable expectations of the indemnitee.[64]

Restitution. In the absence of an express contract, an action for indemnification may be based on a theory of restitution.[65] The position of the *Restatement of Restitution* is as follows:

Title A. Indemnity
§ 76.General Rule

A person who, in whole or in part, has discharged a duty which is owed by him but which as between himself and another should have been discharged by the other, is entitled to indemnity from the other, unless the payor is barred by the wrongful nature of his conduct.[66]

Whether the court finds the equitable claim meritorious will depend on the application of relevant rules, especially the doctrine of active and passive negligence.[67] However, if a contract for indemnification already exists between the parties, it is less clear whether a theory of restitution can support a separate claim for indemnity. More liberal jurisdictions appear willing to review the equitable considerations, irrespective of contract language,[68] while conservative jurisdictions seem to prefer to rely exclusively on the contract terms.[69]

Agency. Under the doctrine of respondeat superior, the employer typically is responsible for the negligent acts of his employees, so long as such omissions occur within the scope of employment.[70] Nothing prevents the employer from asserting a right to indemnification from the employee, though, if the employee's behavior is thought to be inexcusable.[71]

GENERAL CONTRACT LAW AND BUSINESS LAW CONCEPTS

Industrial Sponsor's Exclusive Right to Fund Research in Subject Area in the Agreement

In a contractual arrangement, proposals by the industrial sponsor to exercise control on the manner in which research is being conducted or to expand into subject areas not covered by the agreement should be rejected by the university. However, a sponsoring industry may have legitimate reason for demanding the right to be the exclusive source of funding for a project. The sponsor's goal centers on commercially exploiting the results of the joint research effort, especially when it involves funding a narrowly defined area; the negotiation of a clause establishing exclusive right to use the results of the research is not uncommon. Fowler notes the following illustration:

> For example, the sponsor may request the right to an exclusive license on any patentable inventions which result. In this case it will be important to the sponsor to assure that the results of the work are not "contaminated" with rights vested in others, particularly other industrial or commercial sponsors, but in many cases also including the federal government. In these situations, the sponsor will usually insist on an agreement by the university it will not accept funding from any other source during the course of the funding agreement which could adversely affect its rights to the final result of the research.[72]

Provisions in the agreement granting the exclusive right to fund a project can vary to the extent of control desired by the sponsor. Options include:

> General wording prohibiting the university's acceptance of money during the term of the project if it would alter the rights of industry to that research;
>
> An absolute prohibition against other sponsorship of the same or closely related project without the original sponsor's approval;
>
> An understanding to exclude all additional sponsorship except for those specifically identified parties responsible for some research contribution, such as the federal government or other non-commercial source.

Through these options the industrial sponsor should be able to protect its interests from "contamination" by competing sources.

Technical or Scientific Collaboration by Industry Participants

The nonfinancial contributions by the industrial sponsor may be an essential element in the university's research effort, especially in highly specialized areas or in areas where industry may have unique materials and data essential to the joint study. In situations where the agreement focuses on a cooperative research effort, as opposed to contracted research, the need for industry's contribution should be clearly spelled out to avoid misunderstandings as the project proceeds. Fowler identifies possible problem areas:

> Perhaps some exotic source material (for example, the subject matter of the research) is to be produced or otherwise supplied by the industrial sponsor. When is it to be delivered? In what condition? How will it be transported? Perhaps fabrication or testing of materials, which can better be done in the industrial setting, is to be performed by the sponsor. Who decides when it is to be done and in what quantity? What limitations are there in the industrial sponsor's obligations to produce, deliver, or test? What if the sponsor fails to provide the materials? Will the sponsor provide additional funds to the university to obtain them elsewhere?
>
> Frequently, industry will provide scientists or technicians to collaborate with the university's researchers. The parties should specify the details of such efforts. What is to be the division of labor between the staff? If the industrial researchers are to work on campus with the university researchers, who decides what personnel will be made available, the times they are to be in the campus laboratories, and the number on campus at any one time? It is often best to provide that the university principal investigator has the right to approve which staff will be admitted to university laboratories, as well as their specific schedules, to prevent confusion and unsupervised activities in the labs.[73]

Contractual Reporting Requirements

It may be expected that the industrial sponsor will require scheduled reports of research progress. This requirement should be regarded as distinguishable from any attempt by the sponsor to actually control the research project. Common tests to be applied to reporting requirements in a contract would examine the reasonableness of the requirement, as well as the feasibility and practicality of the reporting form and procedure. A balanced approach between the parties offers the best solution to needs of both organizations. The sponsor may require a monthly technical report which could be burdensome and time consuming, while university researchers may prefer an annual report or no written report. A quarterly or semi-annual report in oral form before a committee might be adequate. Short written summaries of the meeting might provide a satisfactory written document while minimizing bureaucratic reporting. Additional periodic meetings are often recommended to facilitate communication between the parties.

Financial reports accounting for the expenditures of the sponsor's funds should be compatible with the university's financial reporting system. A university's system often runs with at least a thirty-day delay and can appear inflexible when compared with industrial counterparts. Avoiding a customized reporting system avoids the impractical. Because most university sponsored research has historically geared its financial reporting to programs and requirements of federally funded projects, the industrial sponsor may simply want to utilize this reporting procedure for its accounting mechanism.

Funding

Various funding options are available to the parties contracting an industry–university research agreement. In this section of the agreement, the university will be particularly interested in maintaining its customary funding practices, both to expedite payment and to minimize administrative time. Three fund raising options are customarily available for consideration—advanced funding; periodic funding, according to a predetermined schedule of payments; and after-the-fact reimbursement. In some instances, a combination of these options may offer a logical alternative for funding the research project.

Most institutions of higher learning will require some form of advance payment in amounts and at a frequency that will permit a project to continue without interruption. Universities are often hard-pressed for cash, and letters of credit offer a good and customary method for financing a project. Most financial needs can be reasonably estimated in advance of a research venture. Often a detailed schedule of payments with specific dates and dollar amounts will be included in the agreement. This eliminates any uncertainty about future funding and is preferable to requiring the university to make periodic estimates and requests of the sponsor.

Fowler describes two different types of advance funding that might be useful: one method covers the actual costs estimated to be expended in the specific project; the second takes a portion of the funding and treats it as an unrestricted grant within the specified research area:

> For example, an industrial sponsor may agree to provide funds in the amount of $250,000 for each of three years on the following terms: on or before the beginning of each year the sponsor is to pay $100,000 to the university, $50,000 as a grant for the use of researcher X on an unrestricted basis and $50,000 as an advance to be used for the first quarter of the specific research project sponsored under the agreement. Fifty thousand dollars are then to be advanced on or before the first of each of the three following quarters, in each year, all to be used on the sponsored research project.[74]

Lastly, the funding clause of the agreement may be utilized to address the following funding issues:

Overhead costs, if particularly large in amount;

Equipment purchased with research funds shall be owned by one of the parties (typically ownership is retained by the university);

A cost index over a long term project might be utilized to offset the impact of inflation on projected expenses;

Internal distribution of funds if contemplated by the university, might be spelled out clearly to the sponsor;

Construction and remodeling costs essential to the research project should be specified with cost breakdown.

Confidentiality and Publication: the Use of Research Results

The protection of confidential information resulting from research efforts constitutes an essential element in contributing to a corporation's success, growth, and profits. University policies typically lie in the opposite direction. Freedom to publish scientific research lies at the heart of the university's commitment to disseminate knowledge and stimulate scientific inquiry. Contractual negotiations in establishing a research project must reconcile this dichotomy of the university's need to publish with industry's need to protect its proprietary discoveries through patent or trade secret laws. Fowler attributes failures in reaching a sound research agreement to the inability of the industry and university to understand these mutually exclusive concepts:

> This reconciliation is often difficult and sometimes impossible, and this problem probably leads to the majority of cases where industry and the universities just cannot arrive at a research agreement. Few industries today will agree to keep the results of industrially sponsored research confidential. If confidentiality of the results of the research is really important to the industrial sponsor, that firm would be well advised to have the research performed somewhere other than in an academic institution.[75]

The publication of research findings is significant in faculty and researcher recognition and advancement. Through scientific publications a university's research program achieves its stature, success, and reputation. The result of faculty publication efforts probably provides the nexus attracting a particular industry to enter into a specific subject oriented research relationship with a particular university or individual scientist.[76] And, as has been stated earlier, academic freedom concepts support the university's and its faculty's right to publish and disseminate research results.

A legal basis for continued freedom to publish on behalf of the university can be found in the Internal Revenue Code. Organizations exempt under Section 501 (c)(3) of the Code are essentially required to publish their commercially sponsored research that falls within the "in public interest" exemption of Treasury rules and regulations.[77] The right to publish freely and promptly helps establish the necessary substantial relationship between research and the furtherance of the institution's exempt purpose. The goal should be that the research has minimal association with industry's profit base so that it appears as unrelated business income to the tax exempt organization. The right to publish without control helps this research fall under the exclusion from unrelated business income which colleges, universities, and hospitals now enjoy under Section 512(b)(8) of the Code.

For these reasons, non-disclosure agreements are generally not acceptable to the university. These "secrecy" agreements are additionally difficult to administer. Most university administrators are well aware that a campus provides no environment for a secret. Universities cannot guarantee secrecy. Faculty, graduate students, and research assistants typically share their research interests and work with colleagues.

Although a right to publish exists, it is not obligatory. Thus, an option that assists industry in its goal of protecting potential research data, and thus limiting damage to its competitive position, rests on the choice of the university to delay publication for a reasonable time, generally three to six months. Often, it takes four to twelve months for a scientific publication in a journal to reach print. Wilson, Dobb, and Gerl point out in their article the position of Georgia Institute of Technology on this issue:

> Georgia Tech, for its part, is willing in most cases to defer publication for a reasonable period of time. It negotiates the specifics on a case-by-case basis, taking into account four factors:
>
> —the length of time requested by the sponsoring company
> —potential adverse effects of publication on patentability
> —the perceived useful life of the technology involved
> —the extent of the university's interest in publishing information on the technology.[78]

Fowler points out that contractual provisions defining agreement to delay publication vary considerably. Some agreements have the parties merely agreeing not to publish any research results until adequate steps are taken to protect the parties' patent rights. Other more elaborate clauses require pre-publication notice be sent to the sponsor or establish a specific time period during which publication will be withheld, so as to allow time for the filing of patent applications.[79] In the end, however, the decision to publish will rest with the university and the faculty researcher within a specified delay period. The sponsor should not be permitted any control or editorial leeway in textual matters, analysis of scientific results, or statements of conclusions.

Use of Industrial Sponsor's Confidential and Proprietary Information

Many industrial sponsors will provide specific proprietary information and data to the university research team which might be designated as "confidential" or "trade secret." This information is often essential to the research project. The issues involved here are not to be confused with the publication of research or the clauses discussing delay in publication. As was mentioned earlier, the university should avoid any attempt to guarantee its ability to keep secrets since the academic community is founded on the concept of the free exchange of ideas.

Because the campus proves to be a place where confidentiality is difficult to maintain, both parties may question whether they want to establish potentially burdensome controls to protect the sponsor's proprietary information. Other issues regarding liability protection arise if the university researcher secures the proprietary information from someone other than the sponsor. Wilson, Dobb, and Gerl suggest methods that the university can use to avoid the adverse affects that protecting the sponsor's confidential information may have on the free exchange of concepts among faculty and students:

> Some universities limit their obligations, though, by stipulating that "confidential information" does not include the following types of data:
>
> —Information that at the time of disclosure or at a later date is placed in the public domain through issuance of a patent or copyright.
>
> —Information that can be shown to have been independently developed by the university without access to the company's information.
>
> —Data lawfully received from a third party.
>
> —Information disclosed in response to a court order.
>
> In most cases, a university also insists that an R & D agreement designate a time after which the institution is no longer obligated to keep information confidential. And although most universities are willing to negotiate this time period, it is usually set between one and five years.[80]

In final analysis, the inclusion of these confidentiality clauses must balance the needs of the industry sponsor with those of the university. A corporation's legitimate need for secrecy can often be protected for a reasonable time without compromising the need of the university to disseminate research results.

As a special case or example, the use of contracts to impose restrictions on the use of sponsor developed computer software can also present confidentiality problems and directly limit the research effort. Some contracts may limit research rights to specific software or parts of software. The industry sponsor must be careful not to restrict access unduly to data and software that may prohibit further research on the project. University contractors are becoming increasingly aware of the need to differentiate between rights to computer software and patent rights.

Computer software has a complex legal status that is just unfolding. This can present problems to the university administering any contractual software limitations imposed by the sponsor. Although recent court decisions have advanced patent protection for software, copyright and trade-secret approaches are also used to protect innovative computer programs.[81] Copyrighting also has several disadvantages: copyright protects a particular expression of a work and not the idea behind it; it assumes publication, which is undesirable in protecting the source code. It is conceivable that a simple rewriting of a software code by an unauthorized user could circumvent the copyright law. Trade-secret protection discussed later in this chapter may be the best option to prohibit unauthorized use of software. If a program is properly treated as a trade secret, the rewriting of the program does not prohibit legal action for violation of trade secrecy.

Conflicts of Interest

The concept of a "conflict of interest" represents two somewhat related situations—unjust enrichment or conflict in commitment. The classic or traditional definition looks to an unjust enrichment or unfair advantage gained by an individual or institution because of financial or other interests by a party to the transaction. Typically action by the parties improperly influences a business or financial decision by an institution of higher education. The National Association of College and University Business Officers professional code of ethics addresses this type of conflict of interest.[82]

Conflicts of interest or conflicts in commitment are issues that must be addressed individually by each institution of higher education. Since they are primarily ethical/philosophical issues they do not appear appropriate as topics for negotiation in a sponsored research agreement. Fowler identifies one exception to this general rule. This situation involves a continuing consultation, by a researcher with a sponsor, that is a convenient substitute or camouflage for an equity holding position in a company:

> Occasionally, concern over an actual or possible conflict of interest will make it necessary or desirable for the institution to negotiate into a research agreement (or license agreement) a provision whereby the industrial sponsor agrees that no faculty member, staff member (and perhaps student) from the college or university has or will have any interest in, or will participate as an officer, director or consultant in or to, the organization during the course of the agreement.[83]

Certain responsibilities impose the fiduciary duty of undivided loyalty. The essence of fiduciary duty is conveyed in a frequently quoted passage from the famous case of *Meinhard v. Salmon*, in which then Chief Justice Cardozo, writing for a divided court, said,

Many forms of conduct permissible in a workaday world for those acting at arm's length, are forbidden to those bound by fiduciary ties. A trustee is held to something stricter than the morals of the market place. Not honestly alone, but the punctilio of an honor most sensitive, is then the standard of behavior. As to this there has developed a tradition that is unbending and inveterate. . . . Only thus has the level of conduct for fiduciaries been kept at a level higher than that trodden by the crowd.[84]

The definition of a fiduciary is nonetheless uncertain,[85] and the term, or the trust which it represents, is applied rather indiscriminately to public officials,[86] trustees,[87] corporate directors[88] and employees.[89] A common thread unites these classifications insofar as a fiduciary can be said to act "primarily for the benefit of another"[90] and is viewed as a "decisionmaker . . . whose specialized function is . . . recommending or making decisions of a discretionary nature."[91]

The purpose of the fiduciary ethic of loyalty centers on the removal of temptation.[92] In order to further prevent the intrusion of self-interest, conflicts of interest rules have been adopted, especially where the fiduciary has the opportunity to cheat and the power to inflict economic harm.[93] Conflicts of interest rules generally require the full disclosure of any interest that potentially may interfere with fairness and impartial judgment.

Of direct relevance to industry–university research contracts are issues more appropriately defined as "conflicts of commitment." This area of concern was directly addressed by numerous participants in the 1982 hearings held by the House Committee on Science and Technology:

- Representatives Walker and Shanasky expressed concern that Monsanto in its contract with Washington University will acquire access to research funded by federal dollars resulting in giving Monsanto an unfair competitive edge;
- President Derek Bok of Harvard questioned whether scientists in the University should develop an equity interest in corporations involved in comparable research;
- Representative Shanasky raised the conflicts question between a foreign corporation's interests and those of the U.S. government in international contracts such as the Hoechst agreement;
- Attorney Meyerhoff of the National Resources Defense Council suggested a peer review device to screen out questionable equity interest arrangements.

Fowler identifies typical conflicts issues as follows:

1) The concern that the industrial sponsor will attempt to improperly control the scientific or technical approach to the work funded by the sponsor, thereby invading and diminishing the objectivity and independence of the scientific investigator;

2) The problem that faculty investigators, induced by proprietary concerns on the part of the industrial sponsor, may become improperly secretive about their

work, not only to the detriment of free and open dissemination of any scientific and technological developments, but also to the detriment of interaction with and among their students;

3) The concern that the industrial sponsor will improperly attempt to influence or control the choice of, or approach to, future research in the same or related areas. These problems are regarded as acute if the faculty investigators . . . have . . . an equity or some other on-going financial interest in the industrial sponsor.[94]

The duties of the faculty employee are determined by contract,[95] unlike those of the university trustee. But the employee, too, is subject to a duty of loyalty vis-a-vis the employer,[96] although the standard of fidelity is less stringent[97] and perhaps more vague.

If corporate sponsored research is analogous to, or occurs in combination with, "outside" consulting, a conflicts of interest problem may exist.[98] Outside activities per se certainly are not prohibited and, often, are encouraged. Yet, any significant diversion of time and enthusiasm is likely to be viewed unfavorably and ultimately may lead to internal disciplinary proceedings.[99] Because the percentage of faculty members who received consulting fees more than doubled between 1961 and 1974, and the number is highest in the more prominent universities,[100] increasing attention may be devoted to the question of whether such activity undermines the role of the university.[101] Some universities already regulate "outside" consulting as if it presented a conflict of interest problem, imposing a disclosure requirement on faculty members who receive consulting income,[102] which appears comparable to the disclosure requirement for corporate directors.

On the other hand, if corporate sponsored research is but an aspect of traditional academic inquiry, no conflicts of interest problem should exist. Where a research contract is negotiated directly by university and corporate officers, or their subordinates, faculty participation, by definition, is within the university's scope. Even if the research investigator is responsible for attracting corporate funding, so long as the university receives a direct benefit,[103] the resulting agreement would appear to resemble closely government contract research, which the universities have accommodated on a major scale for the past several decades.

A faculty researcher's financial interest in the corporate sponsor may produce a conflicts of interest problem.[104] The argument is not persuasive, except in instances that involve venture capital or an interest in a closely held firm, in which case full disclosure seems appropriate. Shares in a large, publicly owned corporation would not seem to require disclosure; it is improbable that such holdings will adversely affect the university, nor do private for-profit corporations generally require disclosure of their employees' stock holdings. Indeed, such a disclosure requirement might be interpreted as an invasion of privacy.[105]

GENERAL CONTRACT LAW AND STATUTORY GUIDELINES

Intellectual Research and Patent Rights

The two major objectives for higher educational institutions have been identified as follows:

> . . . teaching students from the existing store of knowledge and increasing this store through further research and study. Individual staff members have the implicit obligation to advance either in the arts or the sciences. Thus employment by a university or college carries with it the tacit understanding that the staff member is being paid to teach and to do research and conduct studies.[106]

The objective of research now plays a significant role in an academic institution's formulation of its patent policy as the university seeks to reconcile the interests of the individual investigator, the university, and the corporate body that may be funding the study.

A patent covering any invention or other patentable outcome of research offers a method by which society, through government action, confers upon the holder of the patent certain rights and claims, with respect to use of the invention or concept.[107] The purpose of patent protection is based on the assumption that individuals or society may accrue benefits as a consequence of their invention or discovery. In the offering of a method of protection to the researcher, others are prohibited from taking benefit from a profitable discovery in which they have no interest or rights. Thus, the incentive to pursue research and to develop original and useful ideas is not diminished.

In advocating the use of written patent and copyright policies, the University of North Carolina's Patent Task Force of the University Council for Biotechnology stated:

> Many interests and concerns associated with the typical patent situation center around such questions as the following: Who should own the patent (the inventor, the funding source, the institution, or a combination)? How should rights to use of the patent be shared through licensing provisions? How should the benefits be shared among interested parties through royalties or other provisions? These are just a few illustrative questions. It should be evident from these, however, that agreement among all parties concerned can take many forms. There may be trade-offs of various types. For example, there may be certain income tax laws that make it advantageous for a private party to "own" the patent. But, if a public institution is involved that wants to ensure that its public obligations are fulfilled, ownership by a private party can be counterbalanced by provisions in the agreement that require an appropriate percentage of benefits flow to the public.[108]

A patent has been defined as a "grant made by the government to an inventor, conveying and securing to him the exclusive right to make, use and sell his invention for a term of years."[109] The federal statutory sections of 35 U.S.C. §§ 1 et seq. set forth the requirements and application procedures for obtaining a patent. This same title sets forth the definition of infringement and the statutory mandated remedies. Patent issues and procedures are an integral part of the process of technological innovation. Universities and their industrial counterparts in research must plan to cover patentable inventions in their research contracts so as to protect their royalties and licensing options. Contracting on patent questions between universities and industry typically will reflect a balance between the university's public interests and the industrial sponsor's private interests. Since most institutions of higher education do not have the financial resources or expertise to market patents, they will seek licensees to develop and distribute commercially their patentable research results. The license permits the university to maintain ownership of its property while allowing industrial sponsors to exploit commercially the products. Potential income from profits will benefit both parties.

The issuance of licenses in the area of patents has been traditionally regarded as of societal benefit when the patent holder most likely will not bring the invention into commercial use. In negotiating a license,

> [t]he licensor must state and show unequivocally that the invention contains at least the seeds of sound commercialism; that there are lively expectations of the issuance of a patent; and that there is the likelihood of sufficient profits to justify the commitment of corporate money and effort. The licensor must concede that the invention itself will quite likely need improvement; that the market must be developed and the product readied and distributed to it; and that the license is a contract mutually agreed upon. The licensee must state and show his firm desire, intention, and ability to develop and exploit the market; his financial and entrepreneurial strength; and his willingness to undertake with imaginative thinking the necessary development of the invention itself. He must concede that many difficulties and weaknesses must be overcome, and he must be willing to cooperate with the inventor and licensor in solving these problems.[110]

One of the reasons that educational institutions decide to patent an invention is to encourage the free exchange of information. Often, however, the only way to commercialize the invention is to grant an exclusive license to a specific corporation which will infuse the necessary time and money into bringing the product to market. At this point, the institution must look to the public benefit of the invention and decide whether the open use of the knowledge within the patent outweighs the development and marketing of the specific product.

> In the area of university and industry interplay wherein an invention is made at a university, which admittedly has little interest or expertise in marketing an inven-

tion, the ability to grant an exclusive license is particularly important, again because of the risk involved. Here the commercial developer does not have the advantage of having the inventor in his employ, but at least he has been able to study the invention and the possible market ability thereof before entering into the license agreement. The question remains as to why he should enter into any license agreement which does not give him exclusivity and an incentive to develop the commercial interest. Certainly, if the patent development will involve high costs and considerable risks to the commercial outcome, the potential developer will be willing to proceed only if an exclusive right is granted. Upon commercialization, the university runs the risk of being criticized if the private developer makes a considerable profit from developing the invention. The university shares these profits somewhat through royalties on the patent. The ultimate check on large profits, of course is the incentive they provide for other firms to invent around the patent and enter the market with even newer technology. This is of even greater benefit to the public. The primary interest is one of diligent development, for that is where the public really benefits.[111]

When a university grants an exclusive license, it is essential that a provision be included in the licensing agreement which causes the license to revert to a non-exclusive license if the licensee fails to pursue "vigorously" commercialization or that some other penalty is imposed for failure to pursue commercialization.

University Patent Policy

Due to the complexity of patent application and administration, it behooves an institution to create a usable patent policy well in advance of the first invention. The nature of research is such that a patentable invention may develop at any time, and the prudent administration should be ready for such an eventuality. An example of a mishandled patent opportunity is illustrated by the *Gatorade Case* in which Dr. James Robert Cade in 1965 offered the University of Florida patent rights to his formula for a beverage beneficial to athletes, sensing its apparent commercial value. Dr. Vincent Learned, the Head of Sponsored Research turned down the patent opportunity stating that the university was not interested. Dr. Cade marketed the product initially on his own; it has never been patented. Once marketing was assumed by Stokely Van Camp in 1969, the University of Florida demanded all royalty payments. Suit was initiated in 1971.Total estimated cost of the litigation which was eventually settled is placed at $450,000.[112]

The university's policy should clarify the source of the institution's interest in an invention; identify the rights of the institution, the inventor(s), and the research sponsors; create a method of royalty distribution among interested parties;[113] and establish dispute resolution procedures among the interested parties.[114] Implementation of the patent policy should be carried out by administrators on recommendation of a patent committee composed of faculty members from the arts, as well as science and engineering.[115]

The University of Florida Patent Policy states the rationale for a patent policy in the following manner:

Research, one of the basic objectives of a university, is undertaken to educate students, to stimulate a spirit of inquiry, to solve problems, and to discover new knowledge. Many novel inventions result from research, some of which have the potential for the betterment of society. The University of Florida believes that a university has the obligation to serve the public interest by insuring that such inventions are developed to the point of maximum utilization and prompt availability to the public. University Patent Policy is designed to encourage the creation of inventions by giving adequate recognition and incentive to potential inventors. In sharing the proceeds of royalty-bearing licenses with inventors, the university recognizes inventorship and acknowledges the sizable amount of time and effort necessary to adequately disclose the invention, participate in its evaluation, assist attorneys involved in filing patent applications, and alert potential or actual licenses. It is appropriate and desirable that the University of Florida share in the proceeds of any invention, not only to pay costs of the patent program, but also to support selected University of Florida research programs in recognition of the University's investment in facilities and personnel, without which such inventions would not have been possible.[116]

Florida's state administrative code incorporates all rules governing university employee patents under 6Cl–7.392:

(a) An employee shall disclose all patentable inventions and technological developments which the employee may develop or discover while an employee of the University. With respect to discoveries or inventions made during the course of approved outside employment, the employee may delay such disclosure, when necessary to protect the outside employer's interests, until the decision has been made whether to seek a patent. . . .

(c) Except for discoveries or inventions made during the course of approved outside employment pursuant to a consulting agreement requiring waiver of the employee's rights to any patentable inventions or discoveries which arise during the course of such outside employment, a discovery or invention which is made in the field in which the investigator is employed by the University or by using University funds, facilities, materials, equipment, personnel or proprietary technological information, is the property of the University and the inventor shall share in the proceeds therefrom. . . .

(e) The inventor shall report to the President the nature of the discovery or invention together with an outline of the project and the conditions under which it was done. If the University wishes to assert its interest in the patent, the President shall inform the employee within 20 days after receipt of the report.

(f) The President or his/her designee shall have the power and duty to conduct an investigation as provided in 6C–9.15(2)(e) F.A.C. relating to reporting procedures in patent matters which shall assess the respective equities of the inventor and the University in the invention or technological development, and determine its importance and the extent to which the University should be involved in its protection, development, and promotion.[117]

"Patent Right in Inventions Made with Federal Assistance" or Uniform Patent Legislation

As a part of the Patent and Trademark Law Amendments Act (P.L. 96–517), a new chapter was added to Title 35 and signed into law by President Jimmy Carter in December 1980.This statute has special importance for educational institutions that carry on research with the assistance of federal government sponsorship.

The Uniform Patent Legislation Act was enacted to be effective on July 1, 1981, giving nonprofit organizations and small businesses, with limited exceptions, a right of first refusal to title in inventions they have made in performance of Government grants and contracts.[118]

The Administrative regulations issued pursuant to the law took effect on March 1, 1982, and are applicable to all funding agreements with small business firms and domestic nonprofit organizations executed on or after that date. The terms of P.L. 96–517 and the regulations are not limited to research conducted in the United States, its possessions, or Puerto Rico. They are limited to research carried out anywhere in the world pursuant to funding agreements with small businesses and nonprofit organizations established pursuant to the laws of the United States.

35 U.S.C. § 200: Policy and Objective. This section of the legislation puts Article I, Section 8 of the United States Constitution into practice by "us[ing] the patent system to promote the utilization of inventions arising from federally supported research or development."[119] It would appear from this section that the objective of this legislation is to strike a balance between the rights of nonprofit organizations or small businesses, the Federal government, and industries which will develop inventions by

ensur[ing] that inventions made by nonprofit organizations and small business firms are used in a manner to promote free competition and enterprise; to promote commercialization and public availability of inventions made in the United States industry and labor; and to ensure that the Government obtains sufficient rights in federally supported inventions to meet the needs of the Government and protect the public against non-use or unreasonable use of inventions.[120]

35 U.S.C. § 201: Definitions. This section of the Act defines pertinent terms such as "Federal agency," "contractor," and "invention" in a rather obvious manner, but Arthur A. Smith, Jr., General Counsel for the Office of Sponsored Programs at the Massachusetts Institute of Technology, has observed:

The term "funding agreement" is defined to mean "any contract, grant or collaborative agreement entered into between any federal agency, other than the Tennessee Valley Authority, and any contractor. . . . " This term would seem to quite clearly encompass all of the existing mechanisms for funding support. It is interesting to note, however, that the agencies preparing the regulations implementing this Act appear to distinguish between regulations designed to implement the Act under what is termed "procurement agreements" and those under what is defined as "assistance agreements." In other words, there remains a tendency to continue the artificial distinction between contracts or procurements, and grants or assistance agreements.[121]

The term "practical application," by definition, has the requirement "that the invention is being utilized and that its benefits are to the extent permitted by law or Government regulations available to the public on reasonable terms."[122]

35 U.S.C. § 202: Disposition of Rights. This section has been defined as the heart of the Uniform Patent Act. It clearly states that the nonprofit organization or small business may "elect to retain title to any subject invention."[123] Prior to passage of these amendments, the practice was for universities to negotiate for title with the sponsoring agency, often creating the need for waivers.[124] 35 U.S.C. § 202(a) enumerates three exceptions to the retention of title by the nonprofit organization which are (i) when the funding agreement is for operation of a Government-owned facility, (ii) when the policy and objectives stated in 35 U.S.C. § 200 can be better served by the agency retention of title to the subject invention, and (iii) when government retention of title to the subject invention is necessary for security reasons.

35 U.S.C. § 202(c) lists provisions to be included in the funding agreement. In addition to prompt disclosure of the subject invention and prompt election to retain title, this section grants the Federal agency a nonexclusive, nontransferable, irrevocable, paid-up license to the invention. Section 202(c) also requires a statement within the specification of the patent application that the invention was made with Government support and prohibits assignment of the invention without agency approval, unless such assignment is to an invention management agency.

35 U.S.C. § 203: March-in Rights. This section allows the sponsoring agency either to require the contractor to grant a license or, if the contractor refuses, to grant a license to the subject inventor, if reasonable steps have not been taken to achieve practical application, if necessary for health and safety reasons, if necessary to meet public use requirements, or if 35 U.S.C. § 204 has not been met. The march-in regulations provide:

(1) an informal and rapid agency decision process for determining if the agency shall commence a formal march-in proceeding;(2) a declaration that the agency can terminate a march-in proceeding at any time if it determines a march-in is not

warranted, even though a third party claimant objects to such termination; (3) for closing march-in proceedings to the public when confidential information might be disclosed; (4) that a determination must be issued within 90 days from the completion of the fact finding.[125]

35 U.S.C. § 204: Preference for United States Industry. This section precludes exclusive licensing of any subject invention to anyone unless that person agrees that any products resulting from the "subject invention will be manufactured substantially in the United States." The appropriate Federal agency may waive this requirement on an individual basis if a reasonable effort to comply is shown.[126]

35 U.S.C. § 205: Confidentiality. The purpose of this section precludes release of proposal information under the Freedom of Information Act and allows an inventor time to file a patent application prior to such release.[127] Federal agencies are authorized to withhold information disclosing any invention in which the Federal Government owns or may own a right, title, or interest (including a nonexclusive license) for a reasonable time, in order for a patent application to be filed. Furthermore, Federal agencies shall not be required to release copies of any document which is part of an application for patent filed with the United States Patent and Trademark Office or with any foreign patent office.[128]

According to administrative regulations,[129] an agency may disclose such subject inventions under the Freedom of Information Act, at its discretion, after a contractor has elected not to retain title or after the time in which the contractor is required to make an election, if the contractor has not made an election within that time. The publication of materials describing a subject invention is permitted, to the extent that such materials were provided as part of a technical report or other submission of the contractor which was submitted independently of the requirements of the patent rights provisions of the contract.

35 U.S.C. § 206: Uniform Clauses and Regulations. This section authorizes the Office of Federal Procurement Policy, after receiving recommendations from the Office of Science and Technology Policy, to issue implementing regulations for 35 U.S.C. §§ 202–204 and to establish standard funding agreement provisions.

35 U.S.C. § 207: Domestic and Foreign Protection of Federally Owned Inventions. 35 U.S.C. § 207 authorizes each Federal agency to:

(1) apply for, obtain, and maintain patents or other forms of protection in the United States and in foreign countries on inventions in which the Federal Government owns a right, title, or interest;

(2) grant nonexclusive, exclusive, or partially exclusive licenses under federally owned patent applications, patents, or other forms of protection obtained, royalty-free or for royalties or other consideration, and on such terms and conditions,

including the grant to the licensee of the right of enforcement pursuant to the provisions of chapter 29 of this title [35 U.S.C. §§ 281 et seq.] as determined appropriate in the public interest;

(3) undertake all other suitable and necessary steps to protect and administer rights to federally owned inventions on behalf of the Federal Government either directly or through contract; and

(4) transfer custody and administration, in whole or in part, to another Federal agency, of the right, title or interest in any federally owned invention.

Licensing the Unknown—Intellectual Property and Know-How

Many industry–university research agreements regarding patent rights are unusual in that they are agreements formulated in advance of the innovation. This situation is in contrast to one in which a patent is obtained, then licensed to industry by the patent holder. Greater difficulty exists in negotiating contractual provisions to cover unknown future patents than in creating patent license terms for an existing product, where specific details are known. At the pre-invention stage, the university might be better served by negotiating a flexible, yet comprehensive, provision acceptable to both parties. As Fowler states:

> Such a provision can range from a simple agreement to negotiate in good faith any necessary licensing arrangement after the fact of the invention to a relatively complete license agreement, fully negotiated except for a description of the patent or patents involved.[130]

Another issue which the parties should address involves the sponsor's access to intellectual property, related to the technology developed wholly or partially with industry funds. This might include technical and procedural research data which never reaches the publication stage, due to its lack of significance to the researcher. Some universities automatically give a company a royalty-free, non-exclusive license to use internally the technology developed under their research contracts. Such "know-how" licensing of related technology on a non-exclusive basis presents no practical problem for the university, provided the license terms preserve the university's and the researcher's right to freely publish research data.

The policy at Georgia Institute of technology has been summarized by Wilson, Dobb, and Gerl. They note that Georgia Tech prefers to retain rights to the results of research performed under a corporate–university agreement with the sponsor retaining a royalty-free, non-exclusive license to apply resulting technology to its internal operations. If the sponsor insists on the ownership of patent rights to technology, then Georgia Tech will add a special provision to the contract. The provision waives university's ownership of patentable technology in exchange for additional payments based on a percentage of the total contract. The university also includes a clause reserving its rights to pursue identical research or related research for its own internal use or for third parties.[131]

Under licensing concepts the industrial sponsor and university researchers should be cognizant of the specific statutory guidelines provided under P.L. 96–517 (Uniform Patent Legislation) discussed earlier. This particular act encourages nonprofit organizations to title and market inventions developed with the assistance of Government grants and contracts. Should Federal funds be applied in the research agreement, licensing sections of the act may be applicable and are set out below. In addition, corporate confidential and proprietary information may be made available to university researchers. As discussed below, contracts should include statements specifying the limitations and standard of care to be exercised in handling trade secrets.

The following discussion highlights the significant sections relevant to university licensing under P.L. 96–517 (Uniform Patent Legislation).

35 U.S.C. § 208: Regulations Governing Federal Licensing. The Administrator of General Services is given the power to regulate licensing of federally owned inventions by this section.

35 U.S.C. § 209: Restrictions on Licensing of Federally Owned Inventions. A Federal agency may grant exclusive or partially exclusive licenses, based upon the applicant's intention and ability to bring the invention to practical application and public use, unless such a grant will lessen competition, "resulting in undue concentration in any section of the country in any line of commerce to which the technology to be licensed relates," or is in conflict with the antitrust laws.

35 U.S.C. § 210: Precedence of Chapter. This section states:

> This chapter [U.S.C. §§ 200 et seq.] shall take precedence over any other Act which would require a disposition of rights in subject inventions of small business firms or nonprofit organizations contractors in a manner that is inconsistent with this chapter. . . . The Act creating this chapter shall be construed to take precedence over any future Act unless that Act specifically cites this Act and provides that it shall take precedence over this Act.

35 U.S.C. § 211: Relationship to Anti-trust Laws.

> Nothing in this chapter [35 U.S.C. §§ 200 et seq.] shall be deemed to convey to any person immunity from civil or criminal liability, or to create any defenses to actions, under any antitrust law.

Use of "trade secret" legal concepts may be an option to be considered in contracting as universities and industrial sponsors seek to protect and exploit their research results.

A trade secret is defined by the American Law Institute as follows:

> A trade secret may consist of any formula, pattern, device or compilation of information which is used in one's business, and which gives him an opportunity to obtain an advantage over competitors who do not know or use it. It may be a formula for a chemical compound, a process of manufacturing, treating or preserving materials, a pattern for a machine or other device, or a list of customers. . . .
>
> An exact definition of a trade secret is not possible. Some factors to be considered in determining whether given information is one's trade secret are: 1) the extent to which the information is known outside of his business; 2) the extent to which it is known by employees and others involved in his business; 3) the extent of measures taken by him to guard the secrecy of the information; 4) the value of the information to him and to his competitors; 5) the amount of effort or money expended by him in developing the information; 6) the ease or difficulty with which the information could be properly acquired or duplicated by others.[132]

Because of their dedication to the free interchange of information, educational institutions are traditionally unlikely to make use of trade secrets within their own research. But with the proliferation of industry–university cooperation agreements, university researchers may become privy to proprietary information (trade secrets) in performing these agreements.

The following considerations have been identified for favoring the trade secrets approach as opposed to patenting in protecting proprietary information:

> (1) the avoidance of cost in preparing and prosecuting a patent, (2) the avoidance of risk that any patent might be held invalid in litigation, and (3) the possibility that substantial cost may be involved in enforcing the patent against an infringer.[133]

By contrast, the following considerations favor the patent approach:

> (1) the difficulty of keeping the trade secret a true secret, and (2) the likelihood of a competitor's developing the know-how independently.[134]

The university researcher entrusted with a corporate trade secret must understand that the trade secret is useful to the corporation only as long as it is kept secret.

The Impact of Uniform Patent Legislation on Industry–University Cooperation

Prior to the passage of 35 U.S.C. §§ 200–211, and unless industry had clearly funded the whole project, it was necessary to obtain waivers from the federal funding agency prior to granting any license. In addition, if the industry was unwilling to accept a non-exclusive license, further waivers had to be secured from the funding agency before an exclusive license would be granted. In order to facilitate commercialization, many such exclusive licenses were granted to members of the pharmaceutical industry after satisfying National Institutes of Health requirements, even if the licensee had contributed no funds to the project. This procedure recognized the development costs incurred in bringing the product to market.[135] Passage of the Uniform Patent Legislation has made this whole process easier and has, thereby, opened the area of industry–university cooperation.

As Federal funds for basic research are shrinking, educational institutions are turning to industry for funding. The Pajaro Dunes Conference statement recognized the need to look to industrial sponsorship:

> There are several strong motivations for academic institutions and their faculties to seek industry support for research. First, there is a genuine interest in facilitating the transfer of technology—from discovery to use—to contribute to the health and productivity of society; second, there is interest in ongoing dialogue between academia and industry which could improve the level of applied science by close association with industry applications; and, third, academic institutions and their faculty members are feeling particularly hard-pressed financially and see such cooperation with industry as a way of compensating for a small but important part of the support lost from federal sources.[136]

This increase in industry–university cooperation gives an even greater impetus for educational institutions to formulate clear patent policies. Elements of a good patent policy for educational institutions have been identified by T. L. Stam, Patent Officer at the California Institute of Technology:

> A good policy should protect the rights of the institution and of its faculty and staff in inventions, and should permit the institution to assume its responsibility and obligations relative to its sponsors and to the general public. . . . The responsibility to the sponsor is . . . namely, the moral and ethical and contractual obligation to fulfill requirements relating to inventions which are incorporated in terms of the grants and contracts under which the institution accepts funding for research. Secondly, . . . there is a responsibility to the general public to make an effort to have advanced technology developed by such institutions made available to the public in the form of new and improved products.[137]

Patent policy and patent licensing are a major element of the research agreement between a university and a research sponsor. There exists the need to reconcile the university's goal of free interchange of information by prompt publication of research results with the necessity to withhold such information during the time necessary for patent protection to be secured.[138] In the case of foreign patents, this is further complicated since, under U.S. law, the author of a publication has a two year period in which to file for a U.S. patent. Foreign patent laws vary and may require filing prior to publication. In conjunction with the problem of publication versus patent, there exists the problem of what steps are needed to protect the proprictary or trade secret information which is supplied by the industrial sponsor for use by the university researcher. If an institution decides to accept such information, it becomes obligated to keep such information secret and must develop internal controls to maintain that secrecy, a rather unknown quality within the academic community.[139]

In drafting an industry–university research agreement, identification of individuals who will take title to any patents granted should be stated. Most industrial sponsors will probably not require transfer of patent title, but if they do, the university must decide whether loss of control over dissemination of information and development of the invention outweighs the need for sponsorship by the particular corporation or whether another more amenable sponsor may be available. If the educational institution decides to retain title to the invention, a decision must be made as to the kind of license(s) to be granted. Fowler has observed:

> Most universities, given their druthers, would prefer to grant nonexclusive licenses, as opposed to exclusive licenses or patents, because they are more in keeping with the academic concept of free and broad dissemination of newly acquired knowledge. Universities often have specific written policies declaring the non-exclusive license to be the preferred way to license patents.
>
> Particularly in the case of generic research sponsored on a multi-sponsor or consortium basis, the industrial sponsor sometimes will agree in advance to non-exclusive licenses. The decision to accept a non-exclusive license often will depend on the sponsor's competitive position in its field (if dominant or large enough, it may feel able to forego a preferential patent position), the size of its investment in the project, or the importance of the results of the particular research to the future of the sponsor.
>
> The goal of granting free access to innovations created by universities is certainly laudable. However, as the federal government learned over a period of many years, if a firm cannot obtain some measure of protection by preferential licensing it is not likely to spend significant amounts of money to develop and market the type of innovations which typically result from university research. Thus, most universities must be prepared to grant exclusive licenses to the industrial sponsor in appropriate cases if they are to deal effectively with such sponsors.[140]

An exclusive license should be granted only when necessary to encourage prompt development of an invention and only for a term which ensures such development. The licensee should also be required to exercise due diligence in developing and using the patent.[141]

Regarding faculty involvement in patent aspects of industry–university cooperation, three areas must be addressed. The first is the problem of research being chosen on the basis of the sponsor's desires or its benefit from future patentable inventions. The research agreement must be drafted to allow the academic freedom so necessary to successful basic research.

The second is the problem of faculty conflict of interest. As discussed earlier and as can be seen from testimony at a recent House Subcommittee hearing,[142] substantial ownership by the faculty member or the institution can raise questions of conflict of interest. If such is the case, the granting of a non-exclusive license to the corporate sponsor might be the more prudent approach. The third concern related to faculty participation is purely mechanical and deals with any attempt to patent. This area of concern addresses the necessity for maintaining complete documentation within the project in the event the patent is challenged, including the requirement that each record page be witnessed by a peer who is not involved in the project, a procedure not usually practiced within academic research. A researcher who has had extensive experience in both industry and academia has called the practices of academic research recording "a patent attorney's nightmare."[143]

The questions raised in the area of industry–university cooperation are relatively new. The agreements have increased significantly following the Supreme Court decision in *Diamond v. Chakrabarty*.[144] In the next few years, the questions presently being addressed may become trivial, and, surely, new ones will arise, but of utmost importance is an attempt at careful drafting to anticipate present and future problems.

Federal Tax Laws and Research Relations

The subject area of tax aspects of industry–university research is governed by statute and administrative regulations. The three pertinent sections of the Internal Revenue Code are Sections 501(c)(3) and 44F, created by the Economic Recovery Tax Act of 1981 (ERTA). In addition, new regulations implementing Section 44F have been promulgated.[145] Finally, universities are now exploring Research and Development Limited Partnerships as mechanisms for receiving additional research funding under Section 174 of the Code.

Section 501(c)(3).

Section 501(c)(3) exempts from taxation [c]orporations, and any community chest, fund, or foundation, organized and *operated exclusively* for religious, charitable, *scientific*, testing for public safety, *literary* or *educational purposes*, or to foster national or international amateur sports competition (but only if no part of its activities involve the provision of athletic facilities or equipment), or for the prevention of cruelty to children or animals, no part of the net earnings of which inures to the benefit of any private shareholder or individual, no substantial part of the activities of which is carrying on propaganda, or otherwise attempting, to influence legislation, and which does not participate in, or intervene in (including the publishing or distributing of statements), any political campaign on behalf of any candidate for political office. [emphasis added]

The Internal Revenue Code § 170(b)(1)(A)(ii) characterizes an educational institution as having a regularly scheduled curriculum, a regular faculty, and a regularly enrolled student body. Tax attorneys Paul E. Treusch and Norman A. Sugarman enumerate various "educational functions" such as:

—provision of financial or other aid to scholars;
—provision of free or low-cost housing, books or supplies;
—organizations such as a book store, a cafeteria, a law review which trains students in legal research and writing, a high school athletic association for the promotion of interscholastic competition;
—special training courses, even if conducted outside the traditional educational setting;
—public instruction useful to the individual and beneficial to the community;
—"encouraging or engaging in nonpartisan analysis, study or research" if the primary motive is not to carry on a business for profit;
—dissemination of facts permitting an individual to form an independent opinion on a controversial matter.[146]

The same authors address the "scientific" purpose exemption in the statute. They point out that the pertinent regulations deal mostly with research and that the scientific purpose is usually considered jointly with the educational or charitable purpose to form a "hybrid category." They state,

Even before the regulations attempted a separate definition of the term "scientific," case law set the outer perimeters of the term, making it clear that neither a scientific inquiring purpose nor use of the scientific method would assure "scientific" classification. Neither would the fact that the activity would produce data likely to be of use or interest to scientists. . . . It was equally clear before the present regulations that the research must be in the public interest, not pursued just for private benefit, and such public direction must constitute the "primary" rather than merely an "incidental" thrust of the organization. On the other hand, the "scientific" classification was clearly not ruled out simply because the endeavor constituted applied research rather than fundamental nor because of the controversial nature of the subject matter of the research or study.[147]

Further, the "scientific" classification does not necessarily appear to be ruled out when the "scientific" activities transpire in the commercial setting of a for-profit business when certain criteria exist. These criteria might include making the scientific results freely available to the public or routing the profits back into further study and research.

Treusch and Sugarman state further that

"[r]esearch" is recognized as carried on in the public interest, thereby entitling the organization to the "scientific" label under the regulations in any of three ways: (1) when the results (including any patents, copyrights, processes, or formulae resulting from the research) are made available to the public on a nondiscriminatory basis; (2) when the research is performed for the United States or any of its agencies or instrumentalities or for a state or local government; or (3) when the research is directed "toward benefiting the public," such as through aiding in the scientific education of college or university students, in developing scientific data for a disease, in bringing new industry to a community, or in helping to develop or retain existing industry.[148]

Private ownership, control of resulting patents, or other exclusive rights are not precluded by these tests. Often these methods are the only practical and realistic options in which the public can benefit especially if the research is performed for a government entity or directed toward the public good.

Another issue that impacts universities is found in the Internal Revenue Code at 26 U.S.C. § 513(a) and addresses the unrelated trade or business rule:

The term "unrelated trade or business" means, in the case of any organization subject to the tax imposed by Section 511, any trade or business the conduct of which is not substantially related (aside from the need of such organization for income or funds or the use it makes of the profits derived) to the exercise or performance by such organization of its charitable, educational, or other purpose or function constituting the basis for its exemption under Section 501 (or, in the case of an organization described in Section 511(a)(2)(B), to the exercise or performance of any purpose or function described in Section 501(c)(3)),

except that such term does not include any trade or business,

(2) which is carried on, in the case of an organization described in Section 501(c)(3) or in the case of a college or university described in Section 511(a)(2)(B), by the organization primarily for the convenience of its members, students, patients, officers, or employees, or in the case of a local association of employees described in Section 501(c)(4) organized before May 27, 1969, which is the selling by the organization of items of work-related clothes and equipment and items normally sold through vending machines, through food dispensing facilities, or by snack bars, for the convenience of its members at their usual places of employment.

26 U.S.C. § 512(b)(7)-(9) seems to preclude any problem that universities might have with the unrelated business income in creating research agreements with industry. Paragraph (7) excludes all income derived from research for the United States government, any state or state political subdivision. Paragraph (8) excludes all income derived from research performed by a college, university, or hospital for *any person*. Paragraph (9) excludes all income derived from research performed by "an organization operated primarily for purposes of carrying on fundamental research the results of which are freely available to the general public" for *any person*.

Audits by the Internal Revenue Service of institutions of higher education and other non-profit units often characterize revenues as unrelated business income subject to income taxation.[149] The Internal Revenue Service provided universities with an escape hatch under Rev. Rul. 76–296, 1976–2 Cum. Bull. 142.This ruling states that income from commercially sponsored research will not produce taxable unrelated business income, even it if appears that its primary purpose is for commercial application, as long as it is carried on for the purpose of obtaining and disseminating scientific information. The ruling requires timely publication:

> Thus, commercially sponsored scientific research that otherwise qualifies as scientific research under Section 501(c)(3) of the Code and that meets the publication test of the regulations constitutes research carried on in the public interest. To meet the test, the publication must be adequate and timely.[150]

The definition of the meaning of "research" has significant taxation implications for both corporate sponsors and universities. The Code itself does not consistently use the term adding to the confusion. The term has importance to a commercial sponsor because it receives tax credits and deductions for expenditures for "research activities" and for "research experimental expenditures." The non-profit university likewise receives tax exemptions and avoids unrelated business income tax for engaging in public interest scientific research.[151] The definition of research is still an open ended question with the IRS which industry–university contracts must attempt to address to avoid tax liability. The term "research" is not synonymous with "scientific research," "fundamental research," or "applied research." The Treasury Regulations dealing with exempt organizations state the issues as follows:

> The determination as to whether research is "scientific" does not depend on whether such research is classified as "fundamental" or "basic" as contrasted with "applied" or "practical". On the other hand, for purposes of the exclusion from unrelated business taxable income provided by Code Section 512 (b)(9) it is necessary to determine whether the organization is operated primarily for purposes of carrying on "fundamental," as contrasted with "applied" research.[152]

Adequate attention must be paid to clauses in industry–university research agreements to satisfy the research goals of the agreements while protecting against tax liability. It appears clear that any organization's activities found by the IRS to be engaged in ordinary commercial testing as opposed to "research" will not be tax exempt.[153]

Section 44F. The pertinent section of the Economic Recovery Tax Act of 1981 which became 26 U.S.C. § 44F states:

> (e) Credit available with respect to certain basic research by colleges, universities, and certain research organizations.

> (1) In general. 65 percent of any amount paid or incurred by a corporation (as such term is defined in Section 170(e)(4)(D)) to any qualified organization shall be treated as contract research expenses. The preceding sentence shall apply *only if the amount is paid or incurred pursuant to a written research agreement between the corporation and the qualified organization.* [emphasis added]
>
> (2) Qualified organization. For purposes of this subsection, the term "qualified organization" means—
>
> (A) any educational organization which is described in Section 170(b)(1)(A)(ii) and which is an institution of higher education (as defined in Section 3304(f)), and
>
> (B) any other organization which—
>
> (i) is described in Section 501(c)(3) and exempt from tax under Section 501 (a),
>
> (ii) is organized and operated primarily to conduct scientific research, and
>
> (iii) is not a private foundation.
>
> (3) Basic research. The term "basic research" means any original investigation for the advancement of scientific knowledge not having a specific commercial objective, except that such term shall not include—
>
> (A) basic research conducted outside the United States, and
>
> (B) basic research in the social sciences or humanities.

Under the statute, "qualified research" includes inhouse expenditures for wages, supplies, and services directly attributable to conducting research; sixty-five percent of amounts paid for contract research expenses; and sixty-five percent of corporate expenditures for basic research performed by an institution of higher education or other qualified scientific research organization. It is essential that the university and the corporation enter into a written agreement in order for the industry to claim the credit.

In addition to the credit under § 44F, ERTA added § 170(e)(4) to the Code. This section provides for an increased deduction to *manufacturers* of scientific equipment when they contribute *newly manufactured* equipment to educational institutions that qualify as such under § 501(c)(3). The equipment must be used for research, including research training, or experimentation. The manufacturer must donate the property (i.e. receive no money, property, or services in exchange for the equipment).

Regulations. The regulations state that a contract research expense of a "taxpayer is not a qualified research expense if the product or result of the research is intended to be transferred to another in return for license or royalty payments and the taxpayer does not use the product of the research in the taxpayer's trade or business" unless the taxpayer retains "substantial rights in the research." In addition, the regulations require a written agreement which (i) must be entered into prior to performance of the research, (ii) provides that the research be performed on behalf of the taxpayer, and (iii) requires the taxpayer to bear the costs of the research if the research is not successful. The taxpayer need not have exclusive rights to the results. Section 1.44F–5 deals with specific provisions dealing with basic research and requires that the written agreement include a reporting requirement whereby the institution performing the research is to inform the taxpayer "within 90 days after the close of each taxable year of the corporation what amount of funds provided by the corporation pursuant to the agreement were expended on basic research during the taxable year of the corporation." This requirement applies to agreements effective after 30 June 1983 and has no retroactivity requirement for agreements effective prior to that date.[154]

With the advent of Section 44F, it may become more advantageous for universities and industries to create cooperative research agreements. Both parties will have to look to the Internal Revenue Code and to pertinent regulations in creating those agreements.

The university, especially, needs to be careful that it does nothing to jeopardize its Section 501(c)(3) exemption, which means making sure that any cooperative research complies with "educational" or "scientific" purpose. If challenged, the university can rely on the exceptions found in Section 512(b). In any case, the results of the research must be available to the public. The degree of public access has not been specified and must be articulated under future Regulations as authorized by Section 44F.

Research and Development Limited Partnerships. The research and development limited partnerships are investment mechanisms designed to provide universities with a source of risk capital and investors with tax benefits in the form of tax deductions under I.R.C. § 174 for "research and development expenditures" together with long-term capital gain treatment on the proceeds of the investment under I.R.C. §§ 1231 and 741. Attorneys James C. Garahan, James P. Fuller, and Frederick R. Chilton, Jr., point out the advantages to the RDLP format:

> Companies at various stages of development ranging from "start-ups" to established companies have used R&D Limited Partnerships to finance new products. This type of investment is attractive to investors because the early deductions increase the potential rate of return. Companies find it attractive because the associated tax benefits make it easier to raise funds and because the investors bear the risk of loss if the venture is unsuccessful.[155]

The proposed use of the RDLP by universities is discussed in the final chapter of this work as part of the component on the future of industry–university research relationships. The RDLP has the potential of enhancing the capacity of a university in determining the commercial viability of its technology and permits the university to share substantially from profits generated by commercially successful products.[156]

Using the University's Name

The law relating to trademarks can protect the university's name from commercial exploitation. Any research agreement between industry and universities should specify that use of the university name, logo, or seal by the sponsor is prohibited without prior approval by the university. Terms in the agreement might be broadly stated to prohibit use in the sponsor's sales and marketing of any research products commercially produced, including those undertaken by subsidiary corporations. The function of these clauses protects the good name of the institution of higher education from giving any appearance of endorsing a new commercial product.

The Lanham Trademark Act permits registration of the name and insignia of colleges and universities.[157] A rebuttable presumption of the school's exclusive right to use its marks in interstate commerce exists when registration occurs together with five years of consecutive use subsequent to the date of registration.[158] Registration at the federal level does not create additional common law rights but it does offer significant procedural advantages especially regarding the remedies available.[159]

The function of a trademark is "to designate the goods as the product of a particular trader and to protect his goodwill against the sale of another's product as his."[160] Its purpose, therefore, is two-fold: to protect an imitator from reaping the fruits of goodwill sown by the original user and to prevent the deception of the public as to origin and sponsorship of goods and services.[161]

A cause of action for trademark infringement can be brought under either federal or state law. The owner of a registered trademark is protected against the unauthorized use by competitors or imitators under Section 32 of the Lanham Act, 15 U.S.C. § 1114 (1)(a).[162] Section 43 (a), 15 U.S.C. § 825 (a), prohibits unfair trade competition, irrespective of trademark registration.[163] State statutory and common law are substantially congruent with federal law, generally including protection against trademark infringement and such unfair competitive practices as the misappropriation of another's efforts.[164] Some states also have adopted the Uniform Deceptive Trade Practices Act, a consumer protection measure which creates a new substantive private action for misleading trade identification.[165]

A school may also oppose the subsequent registration of a trademark similar to its own.[166] In order to prevail in such litigation, the plaintiff must show (1) the name or indicia allegedly misappropriated by the defendant is unmistakably identified with the plaintiff and (2) an intent by the defendant to trade on plaintiff's good name.[167]

With reference to industry–university research contracts, the use of the name and logo of an institution of higher education without permission by industry involved in a research relationship should be prohibited by appropriate contractual language. The university should seek to prohibit industry from unfairly profiting in the use of a name which gives the appearance of sponsorship to a certain product. Clear language in the contract should state limitations and restrictions for advertising and promotional purposes that apply to the university name, logo, and the names of researchers and staff. In some institutions, this provision may be similar to restrictions and disclaimers required of faculty involved in private consulting activities.

DISPUTE RESOLUTION

Should the research process be called into question or challenged by either party, the contract of agreement should typically provide for a mechanism for dispute resolution. Despite the best intentions of the parties, an individual researcher may jeopardize a research agreement through dishonest research.[168] With faculty frequently transferring to another institution of higher education, questions can arise as to the ownership of research results. In one situation a Harvard medical researcher received financial support and provided faked and fraudulent research data.[169] In another incident a visiting research scientist was accused of taking to Japan a part of a promising new cell line without permission.[170] More specifically, questions can involve noncompliance with obligations of confidentiality and compliance with research or licensing payment requirements.

Usual methods of dispute resolution often involve establishing non-traditional mechanisms specifically stated in the research agreement. This might involve a mutually agreed upon panel of fact finders. Arbitration is also a common alternative with both advantages and disadvantages.[171]

Although dispute resolution through arbitration is a common method of protecting the partners in research relationships, disadvantages do exist. The danger of the compromise decision and the desire to create a "win-win" situation can result in arbitrators splitting the interests of the parties rather than deciding in favor of one or the other. Some issues are best left to the courts, which may be more appropriately qualified to weigh the issues and interests of the parties. Arbitrators are often unfamiliar with legal rules and could, for example, be unqualified to decide on the validity of underlying patent and trademark issues which may be in dispute. Lastly, attorneys who draft research and licensing agreements will be reluctant to prohibit their clients from resorting to the courts on certain issues.

Five other mechanisms that may be employed to settle disputes have been identified by Chicago attorney Ronald B. Coolley. In making reference particularly to contracts which might result in a licensing dispute, he suggests that parties reluctant to use arbitration might prefer to elect settlement by minitrial. The two basic forms of minitrial involve using a neutral third party who acts as arbitrator (binding minitrial) or as a mediator (non-binding minitrial).[172] Coolley's other mechanisms include a summary jury trial[173] using a private judge as referee, review through a voluntary settlement conference, and referral to private dispute organizations such as the Center for Public Resources, Inc., and Endispute, Inc.[174] The procedures identified by Coolley have a uniqueness in that they are very adaptable for disputes involving technology transfer issues.

Should a dispute between research partners be beyond resolution, there exists the possibility that the contract may collapse. Failure of agreements, especially those involving licensing arrangements, is the subject of a study by Australian lawyer Crispin Marsh. Marsh identifies various causes of technology agreements collapsing. Noting that license agreements, as in many contractual arrangements, are designed to last as ongoing relationships, he categorizes failures into five groups:

1. Failures in the evaluation of the technology;
2. Failures in the selection of the partners;
3. Failures in the conduct of the negotiations;
4. Failures in the agreement itself;
5. Failures in the implementation and servicing of the license.[175]

Marsh further establishes seven philosophical points for lawyers and corporate officials to consider in creating a research relationship:

1. A successful licensing relationship is a long-term one.
2. Analyze the technology in great detail before going into it. . . .
3. Apply the same analysis to your proposed partner. Can he provide what you lack.
4. Negotiate fairly. Don't screw proposed partner for short-term gain. Don't destroy the trust. Do ensure that what has been verbally agreed gets put on paper accurately.
5. Seek an agreement that has positive incentives to achieve the desired results.
6. Take care with the implementation and ensure your licensee continues to value your license throughout its life.

Lastly, but not least, structure the license so that if it does fail the parties may disengage cleanly.[176]

CONCLUSION

This chapter has reviewed the major legal issues which industry–university research agreements must address. The next chapter will examine four specific contracts and discuss the contractual language utilized to address the various legal concerns which have just been identified. Genuine legal problems do confront the parties to these research relationships, and universities in particular need to be prepared to discuss these areas when contracting. The *Speck v. North Carolina Dairy Foundation*[177] case is a good example where utilization of a well-planned patent policy by North Carolina State University could have avoided litigation. The University's patent policy stated that the University would establish equitable arrangements with faculty inventors so they might share in the proceeds of any royalties from their inventions. Professor Speck negotiated with the University for a one-time payment of 15 percent of the royalty fund for his "Sweet Acidophilus" milk process. The University later declined the faculty member's request for a share of the royalties even though the Patent Committee Chairman recommended a one-time settlement. The dictum of the appeal court discusses the faculty member's right to recover profits on the concept of unjust enrichment from the University's misuse of confidential information.

The appeals court frequently referred to the University's Patent Policy reflecting the judiciary's continuing reliance on contractual agreements as the source of mutual understanding. On appeal to the Supreme Court of North Carolina, the Court of Appeals was reversed, with the Supreme Court holding that the secret process developed through the research of Dr. Speck belonged to the university absent a written contract by the university to assign its rights.[178]

Translating legal subject areas into contractual language may appear to the non-lawyer academic as so much "boilerplate," but potential problems must be addressed and unanswered legal questions must be anticipated. The contract represents a memorialization of a mutual agreement between parties—an outline of a research relationship and the potential problems which may exist.

Many legal questions for which there are no firm answers await the university entering into research relations with industry. A few examples follow.

—In the area of traditional academic freedom, what stresses are caused by the research relationship? What is the role of a faculty member as a consultant to industry and how does this effect his academic freedom? Can the faculty member expose the university to liability for his negligence as a consultant? If the faculty member gets himself into legal difficulties with a corporate sponsor, how is he protected?

A prudent faculty researcher will have any consulting contract reviewed by university counsel. Certain small east coast pharmaceutical companies have approached physician researchers at large medical schools requesting their participation as a consultant for a fee—usually small, but including annual business meetings in exotic locations. Dazzled by the request, the physician signs on as a consultant only to discover that he has signed an exclusive consulting contract and cannot participate in a subsequent major industry–university agreement.

—How does the university assure faculty researchers that their academic freedom of research choice will be inviolate? It is not uncommon for a Department Chairperson to direct his faculty's research toward income producing research, especially where tenure questions look to the amount of research funds brought into the university.

—How do faculty and departments protect themselves from loss of funds when an individual faculty member develops money generating research and then transfers to another institution?

—What financial limitations are placed on a faculty researcher's being paid by industry? Court cases support a university's limiting outside activities of faculty. New state statutes and administrative regulations increasingly address conflict of interest questions at public universities. The state of Georgia passed a strong conflicts statute applicable to all state employees including university faculty members.[179]

—In patent policies, how do institutions choose to treat the income that research brings to the institution? Is the royalty income shared with individual faculty or with the department or both?

—When the university's name is associated with a large research agreement or commercial activity, will it upset the local business community? Community relations can be strained when a university affiliates with a commercial venture which may disadvantage other area businesses. This may result in accusations that the university is taking unfair advantage of its tax exempt status.

—If the university's name is associated through research and development to a specific product, is the university's potential liability increased for product liability problems?

—Has an unauthorized use by industry of university material occurred when industry develops a marketable drug as the result of scientific exchanges of university-owned cells protected only by mutual trust and gentleman's agreements? Should these gentleman's agreements be replaced by formal contractual arrangements even when commercialization is not originally contemplated? Has a corporation infringed upon a university's theoretically patentable material when it takes only genetic information from university owned cells? These issues were all raised in court proceedings initiated in 1980 by the Hoffmann–La Roche pharmaceutical house against the University of California over cell lines used to produce interferon.[180] Other issues regarding human cells center on the question now being litigated as to what are the patient's rights in his own cells and genes. In a California case, a leukemia patient claims two university medical researchers used his blood to get a patent on a leukemia treatment that may be worth billions of dollars.[181] Universities should now consider contract language that addresses ownership of human tissue which might result in genetic products and processes which render lucrative profits.

These are a few law related issues which could confront a university entering into a research relationship with a commercial sponsor. Some answers can be anticipated in the body of a research agreement. In the end, the true success of these research ventures will rest in the good will of both parties as they seek to fulfill society's needs for new innovative products.

NOTES

1. Robert M. Rosenzweig, *The Research Universities and Their Patrons* (Berkeley: University of California Press, 1982), pp. 53–54.

2. The Pajaro Dunes Conference was financed by The Henry J. Kaiser Family Foundation and was organized by the following university presidents: Donald Kennedy (Stanford), Derek Bok (Harvard), Marvin Goldberger (Caltech), Paul Gray (MIT), and David Saxon (University of California). Corporations represented included Genentech, Syntex, Gillette, Du Pont, Eli Lilly, and Cetus. See William J. Broad, "Pajaro Dunes: The Search for Consensus," *Science* 216 (9 April 1982):155.

3. "College and Industry Leaders' Recommendations on Commercial Use of Research," *Chronicle of Higher Education* 24 (7 April 1982):12.

4. Ibid., p. 13.

5. Donald R. Fowler, "A Study of the Need for and the Impediments to Improved and Novel University–Industry Research Relationships" (Ph.D. dissertation, Claremont Graduate School, 1983), p. 105. See also, Donald R. Fowler, "University–Industry Research Relationships," *Research Management* 27 (January-February 1984):35–41.

6. John W. Wilson, Jr., Richard P. Dobb, and William T. Gerl, "Consider Various Important Factors When Contracting for University R & D," *EDN* (March 31, 1983), p. 298.

7. See, generally, Donald R. Fowler, "University–Industry Research Relationships: The Research Agreement," *Journal of College and University Law* 9 (1982–83):515–532.

8. Ibid., p. 517.

9. Ibid., p. 518.

10. The need to maintain integrity in scientific research is addressed by a recent report of the Association of American Universities. See William H. Danforth et al., *Report of the Association of American Universities Committee on the Integrity of Research* (Washington, D.C.: Association of American Universities, 1983), pp. 1–6.

11. George G. Bogert and George T. Bogert, *The Law of Trusts and Trustees* (St. Paul, Minn.: West Publishing Co., 1971), § 391.

12. William L. Cary and Craig B. Bright, *The Developing Law of Endowment Funds: The Law and the Lore Revisited* (New York: The Ford Foundation, 1974), pp. 25–36.

13. *Stern v. Lucy Webb Hayes Nat. Train. School for Deaconesses and Missionaries*, 381 F. Supp. 1003, 1013 (D.D.C. 1974).

14. George G. Bogert and George T. Bogert, *The Law of Trusts and Trustees*, § 330 and § 481.

15. See *Cal. Corp. Code*, § 5111 and § 5260 (Deering 1979), requiring funds and assets of non-profit universities be utilized for public or charitable purposes. See also, Annot. 16 ALR 2d 1345.

16. *Cal. Corp. Code*, § 5232 and § 5240 (Deering 1979), directing that corporate directors avoid speculation and perform duties in good faith.

17. See also, Treas. Reg. § 1.501 (c) (3)–1(c) (2) (1976).

18. *Bob Jones University v. United States*, 103 S. Ct. 2017, 2028 (1983).

19. *In re Heard*, 6 B.R. 876, 884 (Bankr. Ky. 1980).

20. See Note, "Developments in the Law—Academic Freedom," *Harvard Law Review* 81 (1967):1049–1159, 1050.

21. *Encyclopaedia of the Social Sciences*, 1935 ed., s.v. "Academic Freedom," by A. O. Lovejoy.

22. Ralph F. Fuchs, "Academic Freedom—Its Basic Philosophy, Function, and History," *Law and Contemporary Problems* 28 (1963):431–446, 431.

23. Note, "Academic Freedom Privilege: An Excessive Solution to the Problem of Protecting Confidentiality," *University of Cincinnati Law Review* 51 (1982):326–352, 329–334; Comment, *University of Pennsylvania Law Review* 130 (1982):712–743; Note, "Preventing Unnecessary Intrusions on University Autonomy: A Proposed Academic Freedom Privilege," *California Law Review* 69 (1981):1538–68; Note, "Academic Freedom and Federal Regulation of University Hiring," *Harvard Law Review* 92 (1978):881.

24. *Cooper v. Ross*, 472 F. Supp. 802, 813 (E.D. Ark. 1979).

25. William A. Kaplin, *The Law of Higher Education* (San Francisco: Jossey-Bass, 1978), pp. 87–94, 142.

26. Note, *Case Western Reserve Law Review* 31 (1980–81):608–609; Matthew W. Finkin, "Regulation by Agreement: The Case of Private Higher Education," *Iowa Law Review* 65 (1980):1120–1200.

27. George L. Joughin, ed., *Academic Freedom and Tenure* (Madison: University of Wisconsin, 1969), pp. 34–36.

28. "Academic Freedom" emphasizes that the important question is how the court construes the meaning of "state action." See Note, "Developments in the Law—Academic Freedom," *Harvard Law Review*, pp. 1056–1064.

29. *Dow Chemical v. Allen*, 672 F.2d 1262 (7th Cir. 1982).

30. 672 F.2d at 1265–6.

31. *Regents of the University of California v. Bakke*, 438 U.S. 265 (1978).

32. *Sweezy v. New Hampshire*, 354 U.S. 234 (1957).

33. *Cooper v. Ross*, 472 F. Supp. 802 (E.D. Ark. 1979), ordering the reinstatement of a teacher dismissed because of political views.

34. *Barenblatt v. United States*, 360 U.S. 109 (1959), upholding the contempt conviction of a teacher who refused to answer questions about alleged Communist Party affiliations is cited for the limitations on academic freedom.

35. *Keyishian v. Bd. of Regents of the State of New York*, 385 U.S. 589, 603 (1967).

36. 672 F.2d at 1276.

37. 672 F.2d at 1279.

38. 672 F.2d at 1279.

39. Russell Kirk, "Massive Subsidies and Academic Freedom," *Law and Contemporary Problems* 28 (1963):607–612, 612; see also, Dallin H. Oaks, "A Private University Looks at Government Regulation," *Journal of College and University Law* 4 (1976):1–12; Philip A. Lacovara, "How Far Can the Federal Camel Slip Under the Academic Tent?" *Journal of College and University Law* 4 (1976):223–240.

40. Richard Hofstadter and Walter Metzgar, *The Development of Academic Freedom in the United States* (New York: Columbia University Press, 1955), p. 419.

41. Kirk, "Massive Subsidies and Academic Freedom," p. 607.

42. Fuchs, "Academic Freedom—Its Basic Philosophy, Function, and History," p. 433.

43. Charles V. Kidd, "The Implications of Research Funds for Academic Freedom," *Law and Contemporary Problems* 28 (1963):619.

44. Barbara J. Culliton, "The Hoechst Department at Mass. General," *Science* 216 (11 June 1982):1202.

45. *Martin v. Helstad*, 699 F.2d 387, 391 (7th Cir. 1983).

46. Ibid., p. 397.

47. *Regents of the University of California v. Bakke*, 438 U.S. 265, 312 (1978).

48. *Randall v. Bd. of Higher Education in City of New York*, 78 A.D.2d 516, 432 N.Y.S.2d 164, 165 (1980).

49. *Smith v. Seaboard Coast Line F. Co.*, 639 F.2d 1235 (5th Cir. 1981).

50. American Law Institute, *Restatement of the Law of Restitution* (St. Paul, Minn.: American Law Institute, 1937), § 75, topic 3.

51. *Hill v. Okay Const. Co.*, 252 N.W.2d 107 (1977).

52. Ray J. Aiken, "Legal Liabilities in Higher Education: Their Scope and Management," section I, *Journal of College and University Law* 3 (1976):181.

53. See *New Amsterdam Casualty Co. v. Waller*, 233 N.C. 536, 64 S.E.2d 826 (1951).

54. John F. Adams and John W. Hall, "Legal Liabilities in Higher Education: Their Scope and Management," section II, *Journal of College and University Law* 3 (1976):244.

55. Kaplin, *Law of Higher Education*, p. 82.

56. *Mink v. University of Chicago*, 460 F. Supp. 713 (N.D. Ill. 1978); *Helmrich v. Eli Lilly & Co.*, 455 N.Y.S. 460 (1982), instituting a cross claim by Syracuse University to the drug manufacturer to contribute for original negligence; *Wetherill v. University of Chicago*, 518 F. Supp. 1387 (N.D. Ill. 1981); 548 F. Supp. 66 (1982); 565 F. Supp. 1553 (1983); 570 F. Supp. 1124 (1983). Annot. 2 A.L. R.4th 1091 (1980).

57. For example, Wis. Stat. § 270.58 (1971) (current supp.).

58. Aiken, "Legal Liabilities in Higher Education," p. 178.

59. Ibid., p. 179.

60. *Smith v. Seaboard Coast*, supra.

61. American Law Institute, *Restatement of Restitution*, § 76.

62. *Barnes v. Lone Star Steel Co.*, 642 F.2d 993 (5th Cir. 1981).

63. *McClane v. Sun Oil Co.*, 634 F.2d 855 (5th Cir. 1981).

64. *Agulnick v. American Hospital Supply Corp.*, 507 F. Supp. 135 (D.C. Mass. 1981).

65. *Commercial Union Ins. Co. v. Ford Motor Co.*, 640 F.2d 210 (9th Cir. 1981).

66. American Law Institute, *Restatement of Restitution*, § 76.

67. *Link-Belt Co. v. Star Iron & Steel Co.*, 65 Cal. App. 3d 24, 135 Cal. Rptr. 134 (1976) (dissent reviews doctrine of active/passive negligence).

68. *E. L. White, Inc. v. City of Huntington Beach*, 146 Cal. Rptr. 614, 579 P.2d 503 (1978).

69. *Jones & Laughlin Steel v. Johns-Manville Sales*, 453 F. Supp. 527 (W.D. Pa. 1978).

70. *Brown v. Wichita State Univerisity*, 217 Kan. 279, 540 P.2d 66 (1975); *First Equity Corp. of Florida v. Utah State University*, 544 P.2d 887 (Utah 1975).

71. American Law Institute, *Restatement of the Law Second: Agency 2d* (St. Paul, Minn.: American Law Institute, 1958), § 401, comment d.

72. Fowler, "University–Industry Research Relationships: The Research Agreement," p. 520.

73. Ibid., p. 521.

74. Ibid., p. 522.

75. Ibid., p. 529. See *Energy Resources Corp. v. Porter*, 14 Mass. App. 296, 438 N.E.2d 391 (1982), where corporate officer was found to have diverted corporate opportunity by deceptively assuming a research project with Howard University and utilizing previously acquired confidential information.

76. But see, *Goldberg v. Medtronic, Inc.*, 686 F.2d 1219 (7th Cir. 1982), where researcher agreed not to publish the results of his work prematurely so as to protect proprietary interests of heart pacemaker corporation only to have his confidential information misappropriated by Medtronic.

77. See Treas. Reg. § 1.501 (c)(3)–1(d)(5) (1976); Rev. Rul. 296, 1976–2 Cum. Bull. 142–143. Consuelo L. Kertz, "Tax Exempt Organizations and Commercially Sponsored Scientific Research," *Journal of College and University Law* 9 (1982–1983):72–76.

78. Wilson, Dobb, and Gerl, "Consider Various Important Factors When Contracting for University R & D," p. 300.

79. Fowler, "University–Industry Research Relationships: The Research Agreement," p. 525.

80. Wilson, Dobb, and Gerl, "Consider Various Important Factors When Contracting for University R & D," p. 299.

81. See, generally, Duncan M. Davidson, "Protecting Computer Software: A Comprehensive Analysis," *Jurimetrics Journal* 23 (Summer 1983):337–425.

82. See, generally, Lanora F. Welzenbach, ed., *College and University Business Administration*, 4th ed. (Washington, D.C.: NACUBO, 1982), p. 182.

83. Donald R. Fowler, "Conflicts of Interest," paper presented at NACUA Mid-Winter CLE Workshop, Washington, D.C., March 11–12, 1983, p. 1.

84. *Meinhard v. Salmon*, 249 N.Y. 458, 464, 164 N.E. 545, 546 (1928).

85. Stanley A. Kaplan, "Fiduciary Responsibility in the Management of a Corporation," *Business Lawyer* 31 (1976):886.

86. *City of Coral Gables v. Weksler*, 164 So.2d 260 (Fla. 1964).

87. *Meinhard v. Salmon*, 249 N.Y. 458, 164 N.E. 545, 546 (1928).

88. Kaplan, "Fiduciary Responsibility in the Management of a Corporation," p. 887, mentions "the American Bar Association Committee on Corporation Laws determined to forego use of the term 'fiduciary' in connection with directors, asserting that the term director should be invested with its own specific content of duty and responsibility."

89. *E. I. Du Pont de Nemours Powder Co. v. Masland*, 244 U.S. 100, 102 (1917).

90. *Restatement (Second) of Agency* § 13, comment a (1958).

91. Alison G. Anderson, "Conflicts of Interest: Efficiency, Fairness and Corporate Structure," *UCLA Law Review* 25 (1978):738–795, 757.

92. Austin W. Scott, "The Fiduciary Principle," *California Law Review* 37 (1949):539–555, n. 30.

93. Anderson, "Conflicts of Interest: Efficiency, Fairness and Corporate Structure," p. 740.

94. Donald R. Fowler, "Conflicts of Interest," p. 2.

95. See *Krotkoff v. Goucher College*, 585 F.2d 675 (4th Cir. 1978); Kaplin, *The Law of Higher Education*, pp. 87–91.

96. *E. I. Du Pont de Nemours Powder Co. v. Masland*, 244 U.S. 100, 102 (1917).

97. Scott, "The Fiduciary Principle," p. 451.

98. Michael C. Weston, " 'Outside' Activities of Faculty Members," *Journal of College and University Law* 7 (1980–81):68–77.

99. Robert M. Rosenzweig, "Faculty and Standards of Ethical Conduct," in Sidney Hook, Paul Kurtz, and Miro Todorovich, eds., *The Ethics of Teaching and Scientific Research* (New York: Prometheus Books, 1977), pp. 79–82; *Gross v. University of Tennessee*, 448 F. Supp. 245 (W.D. Tenn. 1978).

100. Carl V. Patton, "Consulting by Faculty Members," *Academe* 66 (1980):181–85.

101. Weston, " 'Outside' Activities of Faculty Members," p. 68; see also, Rosenzweig, "Faculty and Standards of Ethical Conduct," in which the author describes a disciplinary proceeding at Stanford in 1971 which led to a thorough reexamination of grounds for dismissal of tenured faculty. Kaplin, *The Law of Higher Education*, 1980 supp., pp. 53–55, discusses background pressures, such as inflation and declining enrollments, that may lead to increasing instances of faculty dismissal "for cause."

102. Weston, ibid., pp. 75–76.

103. For example, overhead costs payable directly to the university or the donation of high technology equipment which may offer corporate tax benefits, Tax Report column, *Wall Street Journal*, 22 June 1983, p. 1.

104. Fowler, "University–Industry Research Relationships: The Research Agreement," *Journal of College and University Law* 9 (1981–82): 515–532, 532.

105. Clement D. Johnston and Henry A. Dudley, "Conflicts of Interest in Business and Industry," *Federal Bar Journal* 24 (1964):344–350.

106. Asa S. Knowles, *Handbook of College and University Administration: Academic* (New York: McGraw-Hill, 1970), pp. 1–113.

107. The U.S. patent system finds its creation in Article I, Section 8 of the U.S. Constitution: "The Congress shall have power . . . To promote the progress of science and useful arts by securing for limited times to authors and inventors the exclusive right to their respective writings and discoveries." See also, Edwin Mansfield, *Technological Change: An Introduction to a Vital Area of Modern Economics* (New York: W. W. Norton & Co., Inc., 1971), pp. 127–154.

108. University Council on Biotechnology, "Patent Task Group Report," Chapel Hill, N.C., March 1982 (mimeographed), p. 4.

109. Donald S. Chisum, *Patents: A Treatise on the Law of Patentability, Validity and Infringement* (New York: Matthew Bender & Co., 1983), § 3.06[2].

110. Ibid., pp. 1-115–1-116.

111. John R. Kirk, Jr., "Inventions and Patents Resulting From Use of Government Funds: Movement of Government Patent Policy From Diversity Toward

Uniformity (Maybe)," *1980 Patent Law Annual* 18(1980):133. David S. Saxon, "University–Industry Teamwork," *Les Nouvelles* 14 (December 1979): 232–35.

112. Telephone interview with James R. Cade, 27 July 1983. See also, "Whose Drink?" *New York Times*, 15 August 1971, sec. 4, p. 7; "U.S. Sues Gatorade Maker for Its Profits," *New York Times*, 12 August 1971, sec. L, p. 21; "Stokely Says It Seeks Gatorade Adjudication," *Wall Street Journal*, 17 August 1971, p. 27; "Stokely Settles Suit with Group Contesting Its Rights to Gatorade," *Wall Street Journal*, 2 August 1972, p. 5."UF's Patent Policymakers Working to Settle Dispute Over Royalty Rights," *Gainesville Sun*, 30 October 1984, p. 9A.

113. See *Speck v. North Carolina Dairy Foundation, Inc.*, 307 S.E.2d 785 (N.C. App. 1983), rev'd, 319 S.E. 2d 139 (N.C. 1984), in which North Carolina State University professor sues University agency to recover share of royalties for "Sweet Acidophilus" milk process.

114. K. M. Weeks, ed., *Legal Deskbook for Administrators of Independent Colleges and Universities* (Notre Dame, Ind.: Center for Constitutional Studies, 1982), pp. x–18. See also, Phyllis S. Lachs, "University Patent Policy," *Journal of College and University Law* 10 (1983–84):263–292.

115. Asa S. Knowles, *Handbook of College and University Administration: Academic*, pp. 1–121.

116. Division of Sponsored Research, "University of Florida Patent Policy," Gainesville, Fl., September 14, 1982 (mimeographed).

117. *Florida Administrative Code*, Ch. 6Cl § 7.392.

118. Circular No. A–124, Office of Management and Budget. Patents—Small Business Firms and Nonprofit Organizations, 47 Fed. Reg. 7559–7566 (1982).

119. 35 U.S.C. § 200 (as amended 1980).

120. 35 U.S.C. § 200 (as amended 1980).

121. Arthur A. Smith, Jr., "Implications of Uniform Patent Legislation to Colleges and Universities," *Journal of College and University Law* 8 (1981–82):87.

122. 35 U.S.C. § 201(f) (as amended 1980).

123. 35 U.S.C. § 202(a) (as amended 1980) [emphasis added].

124. Smith, "Implications of Uniform Patent Legislation to Colleges and Universities," p. 88.

125. 47 Fed. Reg. 7582 (1982).

126. 35 U.S.C. § 204 (as amended 1980).

127. Smith, "Implications of Uniform Patent Legislation to Colleges and Universities," pp. 93–94.

128. 35 U.S.C. § 205 (as amended 1980).

129. 47 Fed. Reg. 7561 (1982).

130. Fowler, "University–Industry Research Relationships: The Research Agreement," p. 529.

131. Wilson, Dobb, and Gerl, "Consider Various Important Factors When Contracting for University R & D," p. 299.

132. American Law Institute, *Restatement of Torts*, vol. 4 (St. Paul, Minn.: American Law Institute, 1939), § 757.

133. R. A. Givens, ed., *Legal Strategies for Industrial Innovation* (Colorado Springs, Colo.: McGraw-Hill, 1982), p. 110.

134. Ibid., pp. 110–11.

135. Telephone interview with Dr. Josef Fried, Louis Bloch Professor of Chemistry at the University of Chicago (26 July 1983).

136. "Pajaro Dunes Conference: Draft Statement," *Journal of College and University Law* 9 (1982–83):534.

137. T. L. Stam, "Fundamentals of Patent Policy: Patent Policy for Educational Institutions," in *Technology Transfer: University Opportunities and Responsibilities. A Report on the Proceedings of a National Conference on the Management of University Technology Resources* (Cleveland, Ohio: Case Western Reserve University, 1974), pp. 27–28.

138. But see, *Vitamin Technologists, Inc. v. Wisconsin Alumni Research Foundation*, 146 F.2d 941 (9th Cir. 1945), in which Judge Denman reviews the University of Wisconsin's refusal to license its patented vitamin D irradiation process for use in oleo margarine, which could help cure rickets in poor children, due to dairy industry and state legislative pressures.

139. See, generally, Fowler, "University–Industry Research Relationships: The Research Agreement," p. 527.

140. Ibid., p. 527.

141. "Pajaro Dunes Conference: Draft Statement," p. 534.

142. U.S. Congress, House, Commitee on Science and Technology, *University/Industry Cooperation in Biotechnology: Hearing before the Subcommittee on Investigations and Oversights and the Subcommittee on Science, Research, and Technology*, 97th Cong., 2d sess., 1982, pp. 49–81.

143. Telephone interview with Dr. Josef Fried, Louis Bloch Professor of Chemistry at the University of Chicago (July 6, 1983).

144. 447 U.S. 303 (1980). See also, Iver P. Cooper, "The Patent System and the 'New' Biology," *Rutgers Journal of Computers, Technology and the Law* 8 (1980):1–46; Roman Saliwanchik, *Legal Protection for Micro-Biological and Genetic Engineering Inventions* (Reading, Mass.: Addison-Wesley Publishing Co., 1982), p. 68.

145. See proposed amendments to 26 C.F.R. Part 1, 48 Fed. Reg. 2790 (1983).

146. Paul E. Treusch and Norman A. Sugarman, *Tax-Exempt Charitable Organizations*, 2d ed. (Philadelphia: American Law Institute/American Bar Association, 1983), pp. 127–138.

147. Ibid., pp. 152–53.

148. Ibid., p. 157.

149. See *St. Luke's Hospital of Kansas City v. United States*, 494 F. Supp. 85 (W.D. Mo. 1980), in which the IRS saw fit to tax outside pathology tests performed and interpreted by the hospital's pathology department as unrelated business income under §§ 511–513.

150. Rev. Rul. 76–296, 1976–2 C.B. 142.

151. See Consuelo L. Kertz, "Tax Exempt Organizations and Commercially Sponsored Scientific Research," *Journal of College and University Law* 9 (1982–83):69–84.

152. Treas. Reg. §§ 1.501 (c) (3)–1(d)(5)(1).

153. Treas. Reg. § 1.174–2 (a)(1981); Rev. Rul. 65–1, 1965–1 Cum. Bull. 226.

154. See Lawrence J. Arem, "The Benefits of Research and Development Expenses Increased By New Credit," *Taxation for Lawyers* 10 (Janaury-February 1982):208–213, for a detailed and very helpful discussion of the statute. See also,

Robert L. Black, "The Research Credit: R & D in the Post-ERTA Period," *New York University Institute on Federal Taxation* 41 (1983), § 12.

155. James C. Garahan, James P. Fuller, and Frederick R. Chilton, Jr., "Research and Development Limited Partnerships," in *Research and Development Limited Partnerships* (New York: Practicing Law Institute, 1983), p. 11.

156. See Joseph W. Bartlett and James V. Siena, "Research and Development Limited Partnerships as a Device to Exploit University Owned Technology," *Journal of College and University Law* 10 (1983–84):435–454.

157. 15 U.S.C. § 1051 (1976). A school name and insignia may properly be regarded as service marks; also, a trademark is technically separate from a trade name. But such distinctions have no bearing on trademark registration, nor on the resulting benefits. See David A. Anderson, "Licensing of College and University Trademarks," *Journal of College and University Law* 8 (1981–82):100. For methods of protecting trademarks for institutions of higher education see Shela T. Bell and Martin F. Majestic, "Protection and Enforcement of College and University Trademarks," *Journal of College and University Law* 10 (1983–84):65–67.

158. 15 U.S.C. § 1057 (b) (1976).

159. *Armstrong Paint & Varnish Works v. Nu-Enamel Corp.*, 305 U.S. 315 (1938).

160. *United Drug Co. v. T. Rectanus Co.*, 248 U.S. 90, 97 (1918).

161. *Franchised Stores of New York, Inc. v. Thompson-Hudson Co., et al.*, 300 F. 509 (6th Cir. 1924).

162. Section 32 reads:

> Any person who shall, without the consent of the registrant, use in commerce any reproduction . . . or colorable imitation of a registered mark . . . which . . . is likely to cause confusion . . . shall be liable in a civil action by the registrant.

163. Section 43 (a) reads:

> Any person who shall affix . . . a false designation of origin or any false description or representation . . . shall be liable to a civil action by any person . . . who . . . is likely to be damaged by the use of any such false description or representation.

164. Rudolf Callman, *Law of Unfair Competition, Trademarks and Monopolies*, vol. 2,§§ 15.05, 15.09, 15.18 (4th ed., 1985).

165. *Uniform Deceptive Trade Practices Act*, 7A U.L.A. 35, 39 (1968).

166. 15 U.S.C. § 1052 (d). The pertinent language is:"No trademark . . . shall be refused registration unless . . . falsely suggests a connection with institutions."

167. *University of Notre Dame Du Lac v. J. C. Gourmet Food Imports Co.*, 703 F.2d 1372, 217 U.S.P.Q. 505 (Fed. Cir. 1983). See also, *President & Trustees of Colby Col. v. Colby Col.–N.H.,* 508 F.2d 804 (1st Cir. 1975); *Cornell University v. Messing Bakeries*, 135 N.Y.S.2d 101, *modified*, 285 A.D. 490, 138 N.Y.S.2d 280 (1955), aff'd, 309 N.Y. 722, 128 N.E.2d 421, *re-argument denied,* 309 N.Y. 800, 130 N.E.2d 601 (1955); *University of Notre Dame Du Lac v. Twentieth Century-Fox Film*, 15 N.Y.2d 940, 207 N.E.2d 508 (1965); *University of Pittsburgh v. Champion Products,* 686 F.2d 1040 (3d Cir. 1982), *cert. denied,*______U.S.______, 103 S. Ct. 571 (1982). See *University of Pittsburgh v. Champion Products*, Civ. A. No. 81–2167 (W.D. Pa., June 23, 1983).

168. William J. Broad and Nicholas Wade, *Betrayers of the Truth: Fraud and Deceit in the Halls of Science* (New York: Simon and Schuster, Inc., 1982).

169. William J. Broad, "Report Absolves Harvard in Case of Fakery," *Science* 215 (12 February 1982):874–876.

170. Marjorie Sun, "Scientists Settle Cell Line Dispute," *Science* 220 (22 April 1983):393–394.

171. Robert Goldscheider and Michel de Haas, "The Advantages of Arbitration Provisions in License Agreements," *Patent and Trademark Review* 79 (September 1981):347–354.

172. Ronald B. Coolley, "Alternatives to Litigation," *Les Nouvelles* 18 (September 1983):164.

173. Ibid., p. 165.

174. Ibid., p. 167.

175. Crispin Marsh, "When Agreements Fail," *Les Nouvelles* 18 (September 1983):183.

176. Ibid., p. 187.

177. *Speck v. North Carolina Dairy Foundation, Inc.*, 307 S.E.2d 785 (N.C. App. 1983).

178. *Speck v. North Carolina Dairy Foundation, Inc.*, 319 S.E.2d 139 (N.C. 1984).

179. Act of March 29, 1984, 1984 Ga. Laws 1196.

180. Nicholas Wade, "University and Drug Firm Battle Over Billion-Dollar Gene," *Science* 209 (26 September 1980):1492–1494.

181. *Moore v. Regents of the University of California*, Case No. C513755 (Sept. 12, 1984). See also, F. J. Flaherty, "A Brave New World for Biotech Lawyers," *National Law Journal* (8 October 1984), p. 1.

4

The Contractual Elements of Research Agreements

RECENT DEVELOPMENTS IN INDUSTRY–UNIVERSITY RESEARCH RELATIONSHIPS

As discussed earlier, the two decades from 1948 through 1968 were notable from university research perspectives as that period when government-university joint research agreements prospered. In the mid–1960s industry–university partnerships began to emerge in two identifiable categories. The first was the "industrial associates" model, based on the payment by industry of an annual fee for which it received access to the resources of the entire university and received special treatment from specific departments or laboratories.

In the second relationship, university and industry interaction resulted from the federal government's role as a broker. In this category the federal government provided funds for research in fairly broadly defined subject areas. Government sponsored contracts were let to three teams of two members each and terminated after five years.

The basic appeal to industry of joining in with university researchers in these two categorizations was to fill in gaps of fundamental research that had taken a lower priority to industries' need for applied research.[1] As a result, by May 1974 industry had begun actively to look to the university as a logical source of talent for investigation of scientific issues that require fundamental research techniques:

•**1974** Harvard University Medical School and the Monsanto Company enter into a biomedical cancer research agreement funded at $23 million. As part of the agreement to undertake this research project Harvard also received $4 million for its general endowment to support people affiliated with the research project. Harvard discarded its traditional patent policy that "no patents primarily concerned with therapeutics or public health may be taken out . . . except for dedication to the public." During the almost two year period of negotiations between the parties, the standard practice of peer review through faculty committees and public comment was avoided.[2]

•**1974** Massachusetts Institute of Technology Energy Laboratory. Thomas B. Reed studies feasibility of methanol fuels with unsolicited grant of $100,000 from Minnesota oilman John B. Hawley.[3] Accusations were made that termination of this study resulted from outside influences of Exxon and Ford, which had made large grants to the laboratory.

•**1976**—President Derek C. Bok of Harvard University points out the danger of declining federal support for research and development and,calls for Congress, the Executive Branch, and the Association of American Universities to make a sustained effort to attract support from representatives of business and other sectors to support a vigorous national science research program.[4]

•**1978**—The General Services Administration (GSA) amends federal procurement regulations to permit universities to get a larger share of the commercial benefits of federally financed research.[5] The new regulations provide incentive by encouraging federal agencies to allow universities to retain possession and control of their federally financed discoveries and encourages universities to license these discoveries to private industry.

•**1980**—Massachusetts Institute of Technology and Exxon Research and Engineering Company sign a ten year contract valued at seven to eight million dollars for combustion research. If patentable discoveries result, MIT will grant Exxon royalty-free license for commercial development.

•**1981**—The German chemical corporation Hoechst A.G. provides a ten year grant of $70 million to Massachusetts General Hospital for research in genetic engineering.[6] MGH agrees to grant Hoechst exclusive worldwide licenses to any patentable developments that emerge from the company-sponsored research.

•**1982**—Pajaro Dunes Conference. Heads of five major research universities and eleven corporations meet in California to discuss industry–university collaboration in biotechnology research.[7]

•**1982**—Partners in the Research Enterprise: A National Conference on University-Corporate Relations in Science and Technology is held December 14–16, 1982, at the University of Pennsylvania with eight prominent research university presidents as hosts.[8]

Between 1980 and 1982 corporate investment in academic science proliferated at major American research universities. This increasingly appealing industry connection has provided the basic commodity essential to today's research needs—money. With the federal government's decade of declining support for basic research, universities find these alternative sources for research essential to their future. Science writer Barbara J. Culliton recorded the following examples of corporate support:

—Harvard Medical School has recently created a new genetics department with $6 million from the E. I. du Pont de Nemours & Company to be spent over five years. The principal researcher is Philip Leder.

—Rockefeller University has a five year contract valued at $4 million from the Monsanto Company to conduct research on the structure and regulation of plant genes involved in photosynthesis. The principal researcher is Chua Nam-Hai.

—Stanford University's Channing Robertson and University of California-Berkeley's Harvey Blanch each received $1 million over four years from the Center for

Biotechnology Research to support basic research in the development of chemical processes using genetic engineering microorganisms. The Center is not-for-profit but is financed by the for-profit company Engenics. Engenics is funded by capital from six major corporations: Bendix, General Foods, Koppers, Mead, MacLareun Power and Paper, and Elf Technologies.[9]

Unlike the paramount purpose of government sponsored research with universities where furtherance of the public interest is the goal, industry investments in university research are by desire and obligation centered on the ultimate goal of making a profit. An industrial corporation in terms of its legal definition is an economic for-profit institution. The fiduciary duty of the corporation is to serve the economic interests of its investors.[10] When investing in university research, industry often will place its ultimate profit goal ahead of any service interest to the public at large. With the contracts between industry and universities and the resulting shift from public to private investment in research, a new forum is created which is governed by different rules and goals.

Walsh McDermott of the Robert Wood Johnson Foundation, in reviewing the terms and controversy generated by the Massachusetts General Hospital–Hoechst A.G. agreement, identified the need for clear and unambiguous rules:

> We see ourselves as Dr. Faustus with all sorts of evil things surrounding us. "The real question," he said, "is, 'How does one behave with money?' We're not talking about things that are wrong. We need a code of etiquette."[11]

Robert D. Varrin and Diane S. Kukich of the University of Delaware recommend ten management guidelines that need to be addressed by universities to assure successful ties with industry. They recommend their critical guidelines as keys for universities in averting potential problems or for dealing with problems effectively if they occur:

1. Retain publication rights;
2. Retain ownership of all patents;
3. Establish copyright policies for software;
4. Minimize the use of proprietary information in research and do not require graduated students to sign confidentiality agreements;
5. Create research units with faculty and students, and hire full-time researchers to staff such units if necessary;
6. Faculty should not be permitted to consult with sponsors in the sponsored research area;
7. A faculty entrepreneur's company should not be permitted to sponsor his or her research on campus;
8. Beware of international agreements;

9. Share personnel and equipment with industry;
10. Prepare a model research agreement for potential industrial sponsors.[12]

Attorney Charles C. Caldart noted the philosophical foundation which industry–university contracts should address:

Because it impinges on both sides of the dialectic between academic freedom and public need, the growing commercialization of university research raises serious considerations at each stage of the contract between the university and industry investor. At least four major areas of concern deserve consideration here: the impact on the institutional autonomy of the university; the effect on the nature of university research; the likely restrictions on access to research results; and finally, the long-term impact on university policies.[13]

Specifically reviewing the impact of an institutional autonomy, Caldart raised broad based issues to be clarified by a specific contractual arrangement:

To what degree will industry investors be allowed to participate in the internal administration of universities? The very nature of private investment, of course, presupposes that an investor can exercise some control over the nature and direction of the funded research. . . . But to what extent will an investor be able to affect *directly* university decision-making, beyond the limited confines of the specific research in question? In large part, the formal structure of the relationship between the two parties will be determinative. Where the investor and the university retain separate identities and separate affiliations, the funder may be excluded from participation in the internal affairs of the university; where they do not, such participation will be inevitable.[14]

The research contract can provide this structural separation. Most contracts revolve around a single organizational framework and are a statement of the general rules of the relationship. The form and substance of each contract is usually written to satisfy the institution of higher education, but language is generally clear, simple, and brief. The major issues to be considered in designing an agreement have been identified by Preston W. Grounds, Manager of University–Industry Liaison Programs of the Proctor & Gamble Company:

Checklist Areas to Consider in Designing an Agreement

1. **Description of the Work**
2. **The Principal University Investigator**
3. **The Cost of the Research**
4. **Confidentiality**

5. The Right to Publish
This is unequivocal, and it should be so stated. To satisfy both parties' concerns about disclosure of patentable information, a pre-publication review period is usually considered reasonable if that period is short, but allows for preparation of a sound patent application prior to publication. Note that publication may include an oral presentation, particularly if made before a recognized scientific meeting and preceded by an abstract.

6. Patent Rights and Licensing
An agreement leading to the sharing of rights to potential revenues should consider:

a. Filing
Who files in what countries and who funds filing, prosecution, and maintenance can be handled in different ways but needs to be addressed. Again, it should be noted that the absolute novelty requirement of many foreign countries requires filing before any publication is made.

b. Commercialization
The sharing of rights leading to the sharing of potential revenues according to the contribution of both parties and the constraints of the market is to be diligently sought. The framework for such sharing can be established without specific details which frequently cannot be supplied until such time as an invention is made (if indeed that occurs) and its value assessed. Such a framework can take the following form:

Patent Assignment—Determined according to the policy of the university.

License—Option for either exclusive or non-exclusive rights to the sponsor with royalties or other payments to be negotiated. Royalties may depend upon a number of factors which cannot be determined initially; they may depend upon the type of license, whether or not there is joint inventorship that includes both university and industry personnel, and other considerations. Any period of exclusivity should be sufficient to encourage commercialization.

Diligence—Provisions to encourage diligent commercialization can take the form of reasonable prepaid royalties, although a "best efforts" clause with a stated date for commercialization may be sufficient.[15]

Caldart cautions universities in drafting their contractual rules that the prudent university should retain control over all internal decisions. This division reaffirms the separate nature of the research relationship and preserves the institution's academic autonomy. He notes that institutional integrity is possibly in jeopardy in recent agreements which call for a physical presence on campus of industry's representatives:

Research at the Whitehead Institute, for example, will be directed both by MIT and the Institute. Similarly, genetic research at Washington University, under a funding arrangement with the Monsanto Corporation, will be directed both by scientists employed by Monsanto and by those representing the university. In these institutions, the line between the university and its industry funder is far from clearly drawn. A key feature of such industry/university joint ventures is a university affiliation with the funder which extends beyond a simple contract to perform research.[16]

The following sections of this chapter examine four major contracts to further investigate the language and mechanisms which the parties have mutually agreed will serve as their "code of etiquette."

THE MASSACHUSETTS GENERAL HOSPITAL–HOECHST A.G. CONTRACT OF 1980

The Massachusetts General Hospital–Hoechst A. G. agreement of 1980 is significant as one of the largest—$70 million over ten years—of recent contracts signed in the academic-industrial complex. The contract has become generally regarded as a "model" for industry–university agreements. Biochemist Howard M. Goodman approached officials of the large West German chemical corporation with the proposal to create a molecular biology center of approximately one hundred researchers where these talented individuals could investigate challenging problems without financial limitations. The result was the creation of the new Department of Molecular Biology within the Massachusetts General Hospital (MGH).[17]

The Hospital itself is an independent, not-for-profit entity with its own board of trustees and director. As a teaching hospital it is affiliated with Harvard Medical School, with the Hospital's "academic people" usually receiving Harvard faculty appointments. However, the Hospital is not financially obligated to, nor is it governed by, Harvard's rules. Harvard's president, Derek C. Bok, responded to questions regarding the MGH–Hoechst Agreement stressing the Hospital's independence in what could be interpreted as a negative position:

> I prefer not to comment on it. It was signed without our participation. They did the negotiating, and we heard about it after it was substantially completed.[18]

Obtaining copies of the contractual agreement has been difficult. New York attorney William H. Griesar, who negotiated the contract for Hoechst, noted that there were no plans to make its terms public, stressing the confidentiality of "proprietary" information.[19] Pressure for more information regarding this "model" agreement came from the Harvard faculty with their interests in protecting academic freedom, other lawyers who were negotiating with industry, and Representative Albert Gore, Jr., Chairman of the Subcommittee on Investigations and Oversights of the House Committee on Science and Technology. Congressman Gore obtained a copy of the contract in October 1981 after threatening to subpoena Hoechst and Massachusetts General Hospital officials. The testimony of Ronald Lamont-Havers, M.D., Director of Research at Massachusetts

General Hospital, before the Committee reflects the reluctance of the Hospital corporation to release the terms of the agreement with Hoechst:

Dr. Lamont-Havers: I think, Mr. Chairman, I did send to the committee an outline of the MGH–Hoechst agreement. So somewhere in the mass of papers you have, you have it. Perhaps it would be proper to make that a part of the record at this point in the record.

Mr. Walgren: The Chair recognizes Mr. Gore.

Mr. Gore: Thank you very much, Mr. Chairman. I'd like to request from you a copy of the MGH-Hoechst contract, if wc could have that for the record also. Is that possible?

Dr. Lamont-Havers: That I don't believe is possible. I think if one reads the announcement by MGH, an executive summary of the contract is in there.

Mr. Gore: Well, we'd like to see the contract.

Dr. Lamont-Havers: Then I think you should get in touch with the general director of the hospital and the chairman of the board of trustees.

Mr. Gore: We will do so, and a majority of the subcommittee agreeing, we will subpoena [*sic*] it, if necessary, but I can't believe it'll be necessary.

Dr. Lamont-Havers: I'm in no position to say yes or no to that.[20]

Once a copy of the contract was obtained, Gore requested that the U.S. Comptroller General draft a legal opinion on the question of whether the contract permitted Hoechst to gain exclusive license to any research that was partly supported by funds granted by the National Institutes of Health (NIH).

The Comptroller General of the United States responded on 16 October 1981, concerning the commingling of funds:

MGH must account separately for all expenses leading to an invention including the cost of research itself as well as indirect or overhead costs to be able to show that the expenses were paid with funds provided by Hoechst. . . .

Basically, Section 202 (c)(7)(B) prohibits nonprofit organizations from granting exclusive licenses (except to small business firms) for a period in excess of 5 years from first commercial sale or use of the invention or 8 years from the date of the exclusive license, whichever is earlier, unless the Federal agency involved approves a longer license. Thus, as to an invention subject to a Federal funding agreement, Hoechst would only be able to obtain an exclusive license in the United States for the stated time period unless NIH approved a longer period. . . .

In conclusion, before NIH signs another funding agreement with MGH, it should make clear that the Federal monies involved are not used in conjunction with monies provided under the MGH–Hoechst contract. In the alternative, if the Federal monies are to be used with MGH–Hoechst contract monies for research, the contract should be modified to be in accord with any funding agreement NIH signs with MGH.[21]

An examination of the organization and terms of the agreement between Massachusetts General Hospital and Hoechst reveals the characteristics that make this contract unusual and controversial. As an overview, the contract became effective on 14 May 1981; it contains twenty-nine pages of text organized under thirteen headings entitled "Articles"; it concludes with three "exhibits" and three forms of "agreements" which are incorporated by reference in the basic text.

The following abstract and analysis represents an article-by-article summary of the significant aspects of this contract.

The introductory contractual recital simply establishes that the parties have mutually agreed to do basic research for funding as agreed under the terms and conditions defined in the corpus of the document.

Article I: *Definitions*—Ten different terms used throughout the agreement are defined in this section. This includes such phrases as "field of research," "senior investigators," and "patent application."

Article II: *Rights and Obligations of the Company*—The funding amount and method to be instituted by Hoechst is described in the detail necessary for this rather complicated relationship. Section 2.1 indicates the scope of funding to include renovation, design, and equipment for the Department of Molecular Biology's interim housing; funds for design and construction of the new research building to house permanently the Department; equipment for the new building; and funds for the annual operating costs for the Department in carrying out its sponsored research during the contract term. Should equipment or furniture purchased for the Department by Hoechst funds be transferred out of the Department during the contract's term, then MGH will reimburse Hoechst by the fair market value at that time. Section 2.2 establishes the amount and timing of Hoechst's guaranteed funding and any discretionary funding it might provide. Specific guaranteed funding is stated as follows:

- $2,300,000 for design and renovation of interim space;
- $2,500,000 for interim furniture and equipment;
- $10,800,000 to design and construct a 30,000 square foot new research building for exclusive use by the Department;
- $2,000,000 for furniture and equipment for the new building;
- annual total amounts to be paid quarterly as follows for Department operating costs for sponsored research:

Fiscal Year Ending Sept. 30	Guaranteed Funds for Operating Costs (in addition to guaranteed funds for design, furniture and equipment for the Department)
1981	$500,000.00
1982	$3,600,000.00
1983	$3,600,000.00
1984	$6,000,000.00
1985	$6,000,000.00
1986	$6,000,000.00
1987	$6,000,000.00
1988	$6,000,000.00
1989	$6,000,000.00
1990	$6,000,000.00

A reporting system is established where MGH would provide a quarterly statement of actual operating costs to be reconciled against future and past payments. Thus if MGH requires more funds in one quarter a future quarterly payment could be reduced so as to balance the total payments at the end of the annual payment period. Any additional discretionary funding can be provided by Hoechst after consultation with the Director of the Department and MGH. MGH will receive and disburse all funds. The contract states that guaranteed funds are stated in 1981 dollars and shall be adjusted by the formula in the contract to reflect changes in the dollar value between 1981 and the time of actual payment.

An interesting reimbursement or repayment option is specified in Section 2.3. Under these terms MGH will establish the fair market value of equipment and furniture purchased with Hoechst funds which MGH wishes to retain should the agreement be terminated or at the time of the contract's expiration (stated in Section 3.1 to be 30 September 1990). These funds, as well as money for equipment sold, will accrue to Hoechst. MGH also agrees to repay up to half the construction cost of the new research building upon termination or expiration of the agreement.

Article III of the contract establishes 30 September 1990 as the end of the initial term. The intent of one or the other parties not to renew can be made by notifying the other in writing before the end of the second fiscal year prior to its automatic renewal. Otherwise, the agreement will automatically extend itself for five fiscal years under the same conditions and funding levels. A joint Committee will review the performance of the Department at the end of the 1986 fiscal year. If both MGH and Hoechst agree that the Department is not satisfactorily carrying out its purposes, MGH will correct such situations.

Section 3.2 establishes Hoechst's right to fund exclusively all research within the Department. It is specified that MGH will not seek or accept funding for research within the Department from other profit-making entities. In addition, MGH agrees to prohibit any third party including the United States Government from acquiring any rights or equity in any research solely accomplished by the Department and its personnel.

Article IV states the rights and obligations of Massachusetts General Hospital. Primary among these is its obligation to establish, staff, and administer, as a separate entity, the Department to carry out Hoechst's sponsored research. An initial staff of fifty persons are to be hired until the new research facility is built. At that time a total staff of one hundred persons shall be employed and under the administrative control of MGH. MGH is responsible for preparing an annual operation budget for the Department three months prior to the beginning of each fiscal year. Capital expenditures of $50,000 or more shall not be funded by Hoechst.

Section 4.3 establishes the relationship of certain research personnel within the Department. Howard M. Goodman is to be employed for ten years as Director. The Director and Senior Investigators must be members of the MGH staff nominated for membership in the Harvard Medical School faculty and recommended for tenure at the Harvard Medical School. The Director, Senior Investigators, and Department personnel engaged in sponsored research must all sign an MGH "Participation Agreement" attached to the contract as Exhibit D.

Article V is dedicated to outlining the Department's structure and its responsibilities. Research subjects are to be selected and approved by the Director and are to conform to the goals of the Department. Section 5.2 creates a Scientific Advisory Board of six scientists to review the annual report of the Director and independently evaluate the Department's work. Members on the Board shall include two scientists appointed by Hoechst, two by MGH, and two unaffiliated researchers appointed jointly by the contracting parties. Recommendations from the latter two are specified as advisory only.

Section 5.3 specifically defines the faculty duties of the Director and Senior Investigators, namely, to devote their time primarily to research within the Department. They may devote a reasonable amount of time to faculty duties other than research and consulting for non-profit entities provided it does not interfere with research activities in the Department. In addition, MGH rules, regulations, policies, and procedures are to govern. Senior Investigators are to prepare annual reports including reprints of published articles to be included in the Director's report. MGH is to provide access to Departmental members to its professional staff and facilities.

Section 5.7 provides that the Department will annually present a two to three day symposium for invited members of the academic community discussing the Department programs and research. Hoechst may send an unspecified number of employees to this symposium. In addition, the Department Director is instructed to report on research development directly to Hoechst in Frankfurt, Germany, at least three times per year and Senior Investigators may be called on to confer in Germany at least once per year. Section 5.8 specifies that Hoechst, at any one time, may send up to four qualified individuals to work in and be trained by the Department.

Collaborative work is covered in Section 5.9. It states that scientists may freely collaborate with others within the terms of the Participation Agreement. Collaborative research funded partially by Hoechst and by others must take into account Hoechst's interest in exclusive, world-wide licenses and shall be arranged to entitle Hoechst to the most favorable license arrangement which at a minimum is a nonexclusive license. Without prior approval of Hoechst, MGH agrees not to enter into any agreement for research involving Department employees with any other for profit entity.

Article VI addresses property rights and attempts to balance the interest of both parties. The rights of individual scientists within the Department to publish research results in accordance with educational, scientific, and MGH policies will not be infringed. A thirty day submission clause, which has become a standard, is provided as part of this Article whereby early manuscript drafts authored by Department members resulting from sponsored research are submitted by MGH to Hoechst. If Hoechst decides that patentable inventions are disclosed in the drafts, it may then file for patent protection. At the conclusion of thirty days the scientist at his discretion may then submit the manuscript for publication.

Patent rights and procedures are covered in Section 6.2. All inventions of the Department are to be reported to MGH Office of Technology Administration. Patent application, if filed, will be in the name of MGH but Hoechst will pay application expenses. Should Hoechst not be interested in patenting an invention, it will advise MGH within ninety days of receiving notice of the invention. MGH will then be free to dispose of the patent rights as it deems fit. Similarly, if MGH does not wish to file a patent application, it shall notify Hoechst so that it may file in its own name. MGH, under Section 6.3, will grant Hoechst an exclusive worldwide license for the life of the patent with the right to sublicenses wherever possible. Should Hoechst fail within three years to develop commercially an invention licensed or sublicensed to it under the agreement, the license will then be regarded as nonexclusive and could be made available by MGH to other interested parties.

Royalties from license arrangements are specified in Section 6.4. Specifically it is stated that a royalty rate favorable to Hoechst is expected and reflects due consideration for Hoechst's funding of the sponsored research project. Royalty rates to be paid by Hoechst will not exceed fifty percent of the fair commercial royalty rate for such license. Royalties allocated to the Department and laboratory will be used to pay operating expenses of the Department and will reduce Hoechst's guaranteed funding obligations by that amount. Royalty payments may also be allocated to the MGH general research funds by a formula stated in Exhibit F attached to the contract.

Lastly, Article VII establishes a Joint Committee of three individuals from the MGH Board of Trustees and three senior executives of Hoechst who will meet at least annually to oversee implementation of the agreement and to serve as a forum for communications. Article VIII permits assignment of the parties' rights and obligations within certain guaranty provisions that must be mutually acceptable.

Areas of potential concern and controversy exist throughout this contract. It can be classified as an institution (the corporation) to individual (Goodman) agreement since Howard M. Goodman is specifically required to direct the Molecular Biology Department for the first ten years. The fact that Hoechst can specify the director raises other issues of corporate control. For example, the five senior investigators in the Department are to be selected by MGH after consultation with the corporate sponsor. The contract further interferes with academic matters by specifying that the Department's "academic staff" will be regular members of the MGH staff and, as faculty, will be nominated for appointments and tenure at Harvard as is the case in other MGH departments. Additionally, four Hoechst corporate scientists will be permitted to train in the laboratory at one time.

Goodman as Director along with his senior investigators will select research topics thus asserting MGH's control over decisions relating to research projects and academic matters. However, Goodman is required to report to the sponsor on research progress at least three times per year. The six person Scientific Advisory Board serves a peer review function with its recommendations being strictly advisory. As a point of strength in the contract, the Board has a balanced membership with two members appointed from MGH, two appointed by Hoechst, and two unaffiliated members.

For its exclusive investment in the Department, Hoechst receives rights to worldwide licenses for marketing any commercially useful products developed under its sponsorship. MGH will hold ownership in the patents with royalties being allocated among the investigator, the investigator's laboratory, the Department, and MGH. One consequence of exclusive funding is that the Department's scientists are precluded from seeking NIH grants which take them out of the national peer review process and its beneficial feedback.

Hoechst further extends its exclusive claims by prohibiting the commingling of funds or equipment. The Department was isolated on two separate floors of MGH's Jackson Tower until it moved into its separate research building, the Wellman Research Building, in late 1984. Hoechst agrees to purchase every piece of equipment in the Department and maintains the right to fund additional research, thus safeguarding its exclusive claims.

In the area of preserving academic freedom, selection of the research direction is theoretically left to MGH. Departmental faculty are guaranteed the right to publish all research under certain restrictions. Papers which include new patent related information must be submitted to Hoechst thirty days prior to publication. Faculty are free to collaborate with others under certain restrictions. They must receive written permission from the sponsor before starting a project with another for-profit sponsor. If a collaborative effort is entered into, Hoechst is entitled to the most favorable license obtainable for research results which shall be at minimum a non-exclusive license.

As a "model" contract the MGH–Hoechst agreement set the foundation upon which other relationships could be built. Because it is considered controversial and inflexible due to perceived restrictions and research limitations, subsequent research contracts have attempted to address these concerns and to cover other research issues.

THE WASHINGTON UNIVERSITY–MALLINCKRODT, INC. AGREEMENT OF 1981

In 1981 Washington University in Saint Louis negotiated an agreement with the pharmaceutical company Mallinckrodt, Inc., to undertake hybridoma technology research with corporate support totaling $3.88 million. The scope of the research project involves the production of specific antibodies from artificially created cells called hybridomas. The antibodies hold promise for greatly improving diagnostic medicine and clinical treatment of many diseases.[22]

Freelance writer Kathy Liszewski in her study of the Mallinckrodt, Inc., research agreement explained the development of monoclonal antibodies:

> Monoclonal antibodies are produced in the laboratory by fusing a special type of cancer cell (called a myeloma) with a particular antibody cell.
>
> "It's like two bubbles that get close to each other and fuse to become a larger bubble," said Dr. Joseph M. Davie, director of the Monoclonal Antibody Center at the Washington U. Medical School. The resulting bubble-like hybrid cell (or hybridoma) takes the best of each parent cell: the continuous growth of the cancer

cell and the germ-fighting properties of the antibody cell. What results, therefore, is a non-stop supply of made-to-order, completely identical antibodies.

Antibodies make powerful healing and diagnostic tools because each one is highly specific. Just as each human hand possesses distinctive fingerprints, so does each antibody, which is "imprinted" to search out and destroy only one particular foreign substance.[23]

This research relationship appears typical in that the corporate sponsor appears hopeful that funding for basic and clinical research will lead to commercially marketable products and a future return on its investment. As equal partners in a research endeavor, both parties seek to solidify their relationship through a written contractual arrangement. Strengths of both parties should be capitalized for maximum research results.

Edward L. MacCordy, Associate Vice Chancellor for Research at Washington University, observes this mutuality of interest:

Universities are not reaching out to industry, saying, "tell us what to do, and we will assign a scientist to do it." We're saying, "We're conducting a diverse range of research, you have research interests, and let's see if there is a matching of research and a mutual research that will benefit both of us.[24]

Basic research remains the goal of this agreement. Harry E. Rich, group Vice President for Mallinckrodt's Medical Products Group, reemphasized the separation of the corporate goal for profits from the university's need for independent sponsored research:

There hasn't been a conscious program on our part to go down and educate these people on market awareness. In fact, we've discouraged them from making their own market assessment. That's tough enough for people who know everything that's going on.[25]

The contract itself is entitled *Hybridoma Research Agreement* and consists of thirty-one pages of Articles numbered I through XIII. Incorporated documents are attached at the conclusion of the agreement. These include "Project Investigator's Form of Conditions," "Core Laboratory Director's Form of Conditions," and an "Agreement between Washington University and Recipient Scientist and Institution for Distribution of Cell Line and/or Products Therefrom."

The preamble to the Agreement states the mutual goal of the University and Mallinckrodt to expand research in hybridomas, monoclonal antibodies, and their applications. The University's desire to have useful results made widely available to society through trade and commerce can be matched with Mallinckrodt's desire to develop hybridoma, monoclonal antibody technology. The company has personnel, facilities, and manufacturing processes to permit eventual distribution of medical products through trade and commerce. Through this Agreement the parties will be in a position to take advantage of each others' expertise.

Article I: "Definitions" sets out eight words or phrases utilized in the contract with definitions unique to the contract. For example, "proprietary product" is defined as "any product other than Patented Product which is made, used or sold by the Company or by any party acting on behalf of or under license from Company, which incorporates or is made by the use of any Technical Development made by University or jointly by University and Company which is furnished by University to Company pursuant to the terms of this Agreement."

Article II: "Program Responsibilities" lays out the method for selecting research projects. A project investigator has the responsibility of isolating antigens for hybridoma production and for evaluating the specificity of the monoclonol antibodies produced. Three separate laboratories are created. The Central Hybridoma Production Laboratory has responsibility for maintaining cell lines, cloning of hybridomas, and growing hybrids, among various other highly technical duties. The Central Laboratory will screen the hybridoma culture, identify classes of immunoglobulins, and produce reagents for analysis, plus other technical duties. The Diagnostic Evaluation Laboratory will modify and develop monoclonal antibody testing methods for practical utilization in the clinical laboratory, in addition to other similar technical investigations.

Of particular importance to the faculty researcher is Section 2.6, permitting other research activities. As long as the commercial value of the Mallinckrodt project is not duplicated or diminished, an investigator on informing the Advisory Board may accept research support from other public and private agencies, including commercial firms.

The terms of the Agreement remain in effect for three years from 1 September 1981 as covered in Article III: "Term." The parties are obligated to enter into discussions within thirty days after the end of the second year of the contract to determine if the program should continue beyond its three year time frame. Should the program not be renewed, Mallinckrodt may elect to continue support on a project-by-project basis for any project started but not completed during the three year period.

Article IV: "Program Administration" establishes a five member Program Advisory Committee which controls the entire research program. Three members are affiliated with the University medical school and two with the Corporation. Joseph M. Davie, M.D., of the Medical School is named Program Director and Chairman of the Advisory Committee. An Advisory Board meeting is required at least once each calendar quarter to review financial status and progress of active projects among others, and Mallinckrodt may send alternative representatives to the meetings if deemed necessary. The Program Director will keep and distribute a written summary of matters considered. A majority vote of the Committee is required to defer, disapprove, discontinue, or provide supplemental or continuation of support for approved projects. Annual progress reports from each Project Investigator are to be provided at the end of each term year by the Program Director to the Advisory Committee.

Program finances are established in Article V. The total amount provided to the University during the three year term is $3,881,250, according to the following schedule:

during the first year	$ 1,293,750
cumulative through the second year	$ 2,587,500
cumulative through the third year	$ 3,881,250

Quarterly payments are due on or before the first day of each calendar quarter beginning with September 1, 1981, in the following amounts:

$215,625.00 for each of the first four payments

$323.440.00 for each of the next eight payments

$107,807.50 for each of the last four payments

Through negotiation the payment schedule may be adjusted to adequately reflect the actual and anticipated expenditures.

According to Section 5.4 program funding will be allocated in the following research areas:

-Immunology and Autoimmunity area projects	$ 900,000
-Lipoproteins and Atherosclerosis area projects	$ 450,000
-Malignant Disease area projects	$ 450,000
-Blood Clotting Factors area projects	$ 450,000
-Infectious Diseases are projects	$ 450,000
-Shared Costs of Core Laboratories	$ 1,181,250
total authorized funds	$ 3,881,250

It is noteworthy that Section 5.9 specifies that the title to equipment purchased under this agreement with program monies vests in the University.

Section 5.10 directs a final accounting on termination of the agreement and a return to Mallinckrodt of any remaining funds.

Publication procedures and cell line protection are covered in Article VI. Of primary importance to faculty investigators is the assurance that they are at liberty to publish or disclose the results of their research, restricted only in that program participants must provide copies of articles intended for publication to the Advisory Committee at least two weeks before submission to a publisher. The Advisory Committee will screen patent applications for descriptions of inventions that have not been filed or for unauthorized disclosure of Mallinckrodt proprietary information. The corporation by written request can delay publication for up to three months from date of transmittal to the Advisory Committee, allowing the corporation to file for patent protection or seek deletion of its proprietary information.

The corporation agrees in Section 6.3 to review pre-publication articles for potentially patentable inventions promptly. Section 6.4 acknowledges that Mallinckrodt should be fully aware of research progress through continuing communication and will have full opportunity without delay to establish patent rights.

Lastly, Section 6.6 anticipates the distribution of cell lines to other researchers outside the program. Scientists are free to distribute cell lines or products to scientific colleagues provided that: a) they be used only for research purposes, b) they may not be used for commercial purposes or to the benefit of other commercial organizations, and c) they will not be distributed further to other parties.

The need for confidentiality is addressed in Article VII. The first area of protection involves Mallinckrodt's agreeing not to disclose University technical developments as it constitutes valuable University property. The corporation agrees that for ten years it will protect all confidential technical information, although it will not be liable for unauthorized revelations made despite its precautions. For confidential corporate information, it is acknowledged that Mallinckrodt has the responsibility to obtain personal commitments of confidentiality as it deems necessary from University personnel requiring these researchers to keep such information confidential.

Patents and licensing arrangements are described in Article VIII. Mallinckrodt will notify the University when potentially patentable inventions are first developed. In turn, the University will notify Mallinckrodt of potentially patentable inventions. At its cost and initiative Mallinckrodt will file U.S. and foreign patent applications with title being in the University's name. However, as stated in Section 8.11, Mallinckrodt will have the right of first refusal to obtain licenses from the University for inventions resulting from the program. These licenses according to Section 8.12 will be exclusive for the life of the patent. If the invention comes from a joint support project with another sponsor, then Mallinckrodt can request not less than an exclusive license for a limited number of years or a non-exclusive license for the life of the patent. License grants on hybridoma cell lines made by program support, but with no patent application, shall be licensed by Mallinckrodt exclusively for ten years.

Section 8.14 discusses royalty payments based on Mallinckrodt's representation to introduce licensed products in the commercial market place. The corporation is permitted to sublicense to others but will pay royalties to the University. A detailed royalty schedule is stated in Section 8.14(d) 1–5. Section 8.14(f) prints the first indemnification agreement. This provision states that Mallinckrodt will indemnify the university for liability arising from the use of the university's technical developments and from liability arising from the use, sale, or disposition of products made with this technology including products made by sublicensees. This provision survives the termination of the license agreement. Section 8.14(h) states that the law of Missouri governs the contract.

Article IX sets forth standard termination procedures. Article X again restates the corporation's holding the University harmless to suits for liabilities and damages arising out of damages resulting from joint research efforts under the program. This article survives the termination of the Agreement. Article XI provides for transfer of interests if each party mutually agrees in writing. Article XII provides addresses for official notices regarding the contract and Article XIII establishes general provisions and restrictions.

In analyzing this agreement, it can be classified as an institution to institution contract. The significance of an administrative Program Advisory Committee with a membership of five rests in its make-up: four of the members are from the University (Davie as chairman, Lacy, Kipnis, and McDonald) and one is selected by the sponsor (Oesterling). In terms of control, this is certainly a strength for the University since the Advisory Committee selects projects under this program which are in turn carried out under the direction of a project investigator.

This contract's limitations apply in a general way to research restricted to the development of hybridomas and monoclonal antibodies as applied in areas of immunology and autoimmunity, infectious diseases, among others. It also includes research services for production of hybridomas and antibodies by core laboratories. There appears to be a good deal of flexibility in selecting research topics due to the somewhat broad scope permitted by the contract.

In terms of academic freedom and conflicts of interest in research, this contract also seems to favor the University. A project investigator may accept research support from other public or private agencies including commercial firms for research investigations that do not diminish the commercial value of the sponsor's projects. Additionally, title to all equipment purchased with program monies rests in the University.

Publication of research results and cell lines by investigators is permitted with some restrictions. On written request from the sponsor, the University will delay publication for three months from the date of transmittal to the Advisory Committee. This permits the sponsor to file for patent applications. The sponsor may also require the writer to delete corporate proprietary information. All publications are to acknowledge that the research was sponsored and supported by the company. On the positive side researchers may exchange cell lines freely with non-commercial peers and colleagues for non-commercial experiments when those peers sign a written agreement acknowledging the stated restrictions.

This agreement permits the corporate sponsor to secure personal commitments of confidentiality as it deems necessary from University personnel. This is for the purpose of protecting company confidential information and is not a common practice in nonprofit institutions. Royalty payments are divided by the University with no monies being assigned to one individual. Payments are allocated with forty percent to the research laboratory, forty percent to the Department, and twenty percent to the School of Medicine. On the whole, this agreement exhibits a variety of terms which are very favorable to the University.

THE WASHINGTON UNIVERSITY–MONSANTO BIOMEDICAL RESEARCH AGREEMENT OF 1982

Building on the 1981 Mallinckrodt, Inc., agreement, on 3 June 1982 Washington University and Monsanto announced a five year research agreement calling for $23.5 million in support. Similar to the Massachusetts General Hospital–Hoechst A.G. agreement, this contract received widespread publicity as another "model" for research agreements. The Saint Louis press noted this characteristic as follows:

> This agreement has drawn praise both from colleges and corporations for the care it pays to questions that had been raised about similar liaisons—questions about academic freedom, ownership of results of the research and the merging of the separate cultures of the company and the university. . . .
>
> The agreement is considered a model that other institutions could use to shape contracts on their own.[26]

David M. Kipnis, M.D., Busch Professor and Head of the Department of Internal Medicine at Washington University in Saint Louis, highlighted the significance of the process followed in achieving an agreement as a model when testifying before House Subcommittee on Investigations and Oversights of the Committee on Science and Technology:

We think that our agreement represents a significant step in the evolution of industrial-academic collaboration. And although the specifics of the agreement suit the unique traditional geographic and scientific characteristics of Monsanto and Washington University, we feel that some of the institutional values and procedures which we used in evolving this kind of agreement may be useful to other institutions seeking to develop similar collaborative programs with industry.[27]

Howard Schneiderman, Senior Vice President for Research and Development for Monsanto, in his testimony before the Subcommittee stressed the process of reaching an agreement between two differing cultures as the significance of the contract, not that it should serve as a model:

We do not view our agreement for a joint drug discovery venture as a model for other companies and other universities. It was designed to suit our particular institutional cultures. Another pair of institutions would have to design a plan that fits their cultures and their needs. Indeed, the only thing about our contract that I would commend to others is the process by which we put it together.[28]

The scope of the research project covered by the contract is significantly broad. Funding from Monsanto will support research projects in the field of cellular communications which generally involves the study of how cells function individually and with each other. The collaborative effort involves the field of biomedicine with a focus on proteins and peptides that modulate cellular functions. Some possible research areas that fall within this broad scope include the study of drug receptors and inhibitors which determine how drugs work in the body, the study of blood coagulation, research into the body's immune system, and studies related to proteins and hormones that travel between cells.[29]

The contract consists of fifty pages of text arranged under sixteen "Articles." Attached is an "Exhibit A" entitled "Agreement of Program Participants" which consists of seven pages. The contract preamble establishes that the Agreement between Washington University and Monsanto Company becomes effective on July 1, 1982. It recognizes that each party can benefit from a research relationship involving biomedical research over a period of years, that potential benefits to health care consumers could result and are likely to be brought to public use and benefit through the incentive of patent protection, that academic freedom can be preserved in this corporate/university relationship, and that the scope of the collaborative effort will focus on proteins and peptides that modulate cellular function.

Article I: "Purpose and Scope of This Agreement" simply establishes that the purpose of this writing is to recite the contractual framework which will govern the collaborative efforts in multiple research projects to be undertaken as authorized by the Advisory Committee.

Article II: "Definitions" contains eight terms as used throughout the contract. For example, "Program" means "all research activities performed by or for the University under this Agreement which are authorized and funded by the Advisory Committee and Program Director from financial support provided by Monsanto." "Advisory Committee" is subsequently defined in Section 2.3.

Article III: "Term of Agreement" states that the contract will last for five years beginning on July 1, 1982, and terminating on June 30, 1987. Section 3.2 allows for discussion to begin on February 1, 1985, on the question as to whether both parties desire to continue the program beyond the stated termination date. Monsanto will have the option as stated in Section 3.3 to continue any Project not completed during the normal term for not more than an additional two years.

Article IV: "Program Administration" establishes the governing mechanism for this agreement. Section 4.1 indicates that the Advisory Committee of seven members will be chaired by David M. Kipnis, M.D., of the University and will also include Doctors Luis Glaser, Paul Lacy, and Joseph Davie. Monsanto will be represented by five members including Doctors Howard A. Schneiderman, G. Edward Paget, Louis Fernandez, and David C. Tiemeier. Section 4.2 establishes that a majority vote of the Advisory Committee will be required to approve, defer, or disapprove funds for new, supplemental, or continuing projects. Other sections in this article discuss appointment procedures for replacing Board members, requirements for quarterly meetings, distribution of written summaries of matters considered and actions taken by the Board, and provision for alternates to attend the Board meetings.

Article V: "Project Selection and Implementation" establishes the role of the Advisory Committee in selecting all exploratory and specialty projects as well as having responsibility for the overall direction of the program. The Advisory Committee identifies fields of interest, seeks project proposals from university faculty members, selects the project investigator, defines the research activity to be pursued, establishes a budget for research, and defines the time duration for the project which must be consistent with the time limitations of the contract.

Section 5.2 sets forth an unusual project selection guide not found in most industry–university research agreements. It states that the parties expect the research programs to embrace two types of projects in a specific ratio—30 percent of the research effort will be directed toward fundamental exploratory projects (basic research) and 70 percent will be directed toward specific products or specialty projects (applied research).

With the approval of the Advisory Committee, Monsanto may designate a Monsanto Project Scientist to serve as the primary contact with the project investigator. The Program Director submits to Monsanto a written summary report of all important findings and results as soon as available in addition to a detailed annual report on the anniversary of the contract.

Article VI: "Interaction between Monsanto and the University" clarifies the involvement of Monsanto in the various projects. Section 6.1 requires the University to ensure that its scientists will be available for consultation with appropriate Monsanto scientists and that the University will provide temporary office space for collaborating Monsanto scientists. Section 6.2 permits Monsanto scientists and technicians to spend time in University project laboratories to learn the techniques being developed and to facilitate technical information transfer. Section 6.3 anticipates cooperative interaction between Washington University and Monsanto scientists in identifying facilities of Monsanto which might be used by the University researchers to advance the research projects. This cooperative spirit is also expected in evaluation of commercial potential of research results. Unlike the MGH agreement, the Monsanto agreement places no limit on the number of Monsanto scientists who may utilize the University's laboratories.

Article VII: "Scientific Review Panel" determines the scientific merit and cost effectiveness of projects by periodic review of an independent panel of scientists. The panel consists of four distinguished scientists not affiliated with either party who, in the third year of the contract and every two years afterward, will review and appraise the program both qualitatively and quantitatively. Their confidential report will be submitted to the Advisory Committee, the University Chancellor, and the Chief Executive Officer of Monsanto.

Section VIII: "Program Finances" provides the funding guidelines for the agreement. Monsanto will pay $23,500,000 over the five year term of the contract with an inflation adjustment formula provided in Section 8.2.The schedule of payment is allocated as follows:

Contract Year	Exploratory Projects	Specialty Projects	Construction and Renovation Projects	Contract Year Budget
82/83	$ 1,500,000	$ 1,500,000	$ (See Para. 8.4)	$ 3,000,000
83/84	$ 1,600,000	$ 2,200,000	$	$ 3,800,000
84/85	$ 1,700,000	$ 3,000,000	$	$ 4,700,000
85/86	$ 1,800,000	$ 3,800,000	$	$ 5,600,000
86/87	$ 1,900,000	$ 4,500,000	$	$ 6,400,000
Total	$ 8,500,000	$ 15,000,000	$	$ 23,500,000

The Program Director will allot funds with the approval of the Advisory Committee and may make adjustments among expense categories if justified by the project investigator. Among a variety of controls, Monsanto has access to all accounting records for audit by either its auditors or outside auditors. The University will bill Monsanto monthly for project expenditures, and Monsanto will pay promptly upon receipt of invoices. Section 8.10 permits the title to all equipment purchased with program funds to vest in the University at time of purchase.

Article IX: "Publications and Review of Technical Developments" specifies that scientists will submit copies of abstracts and articles of proposed publications to a Monsanto member of the Advisory Committee at least one month prior to submission to a publisher or third party. If Monsanto upon prompt review determines that potentially patentable developments are disclosed, it will immediately inform the University that a delay for a reasonably brief time will be required to establish patent right. This section, therefore, states that technical developments may not be published without the agreement of Monsanto. When read with Section 10.2 below, this section indicates that these technical developments will normally remain not disclosed (secret) for a minimum of two years. For verbal presentations scientists are provided guidance from both the University and Monsanto to avoid divulgence that would compromise any patent efforts. Monsanto acknowledges that its project scientists shall cooperate in seeking immediate action to establish its patent rights and to avoid delay in filing appropriate patent applications. Section 9.6 states that research samples are also subject to confidentiality, patent and license protection and that Monsanto has a right to receive and use all samples.

Article X: "Confidentiality" specifies that all technical developments and patents are the sole and exclusive property of Washington University subject to an initial license right in Monsanto. Monsanto retains access to the research information, but it owns neither the raw data nor the ultimate patents. Section 10.2 requires Monsanto to safeguard confidential technical information and not to disclose this data for an initial two year period which may be extended at the University's request for another two years. Any liability of Monsanto, however, for unauthorized disclosure of technical developments is specifically waived. Section 10.3 further limits Monsanto's confidentiality obligations by specifically not extending its obligations to technical developments

—that are part of the public domain,

—that were in Monsanto's possession prior to disclosure by the University,

—that were received by Monsanto lawfully from a third party, or

—that have been revealed in patent applications.

Section 10.4 addresses the obligation of the University in handling Monsanto's confidential information by University personnel. This section provides that program participants sign a form for a personal commitment of confidentiality in advance of receiving confidential information. These personal commitments of confidentiality are standard practice in the corporate sector where proprietary information must be protected, though not a usual practice in universities. The University clearly states its inability to make commitments of confidentiality on behalf of its faculty and states its inability to control confidential information disclosed to its faculty. This waiver creates a confidentiality relationship directly between an individual faculty member and Monsanto as stated in the terms of any signed personal commitments of confidentiality.

Article XI: "Patents and Licensing" occupies seventeen pages of the contract and consists of the largest and most detailed article within the agreement. The more significant sections specify the following duties:

11.1—Monsanto has no obligation to carry out commercialization;

11.2—Monsanto is obligated to monitor research progress through its representatives to facilitate easy detection of patentable inventions;

11.5—Monsanto's interest in licensing a prospective patent to an inventor will require its patent attorneys to file the U.S. Patent Office application and corresponding foreign patents;

11.8—Washington University will retain title to all patents concerning inventions and technical developments made under this agreement. The University will, however, consider the relative contributions of the co-inventors in establishing royalty rates with Monsanto;

11.11 and 11.12—The University and each inventor waive any claim against Monsanto for injury, loss, or damage that might result from acts of omission or commission in preparation and filing of the patent applications;

11.14—Monsanto is granted permission to license and sublicense to manufacture and market inventions in other countries which it may elect;

11.16—Monsanto is awarded an irrevocable license to all technical developments not covered by patents. Monsanto agrees to indemnify the University for any liability which might result from use of the technical developments;

11.18—Monsanto is granted licenses under stated terms and conditions. Monsanto is required to make reasonable and early efforts to introduce licensed products into the marketplace. A product development plan with a schedule of steps for market entry will be provided to the University. If a sublicense is awarded to others by Monsanto, it must notify the University. Royalty rates will be negotiated in good faith and will recognize Monsanto's contribution to the research. Monsanto will indemnify the University for liability arising from any use or sale of licensed products;

11.19—Monsanto's royalty formula entitles it to a credit for certain expenses not to exceed 25 percent of gross royalties commencing with the fourth year in which royalties are due.

11.20—Permits the University to offer to others those licenses which Monsanto does not wish to market.

Article XII: "Termination" is stated as effective on June 30, 1987, unless extended by mutual agreement. Termination may also be effected by default, breach of provisions, insolvency, assignment of royalties to creditors, or bankruptcy.

Article XIII: "Indemnification" provides for Monsanto to hold harmless, indemnify, and defend the University from all liabilities and costs which may develop from use or sale of products resulting from this agreement. University research personnel are covered by appropriate insurance and the University will not hold Monsanto liable for personnel injuries or loss of life incurred in the performance of the research work not attributable to Monsanto's negligence.

Article XIV: "Transfer of Interest" is prohibited without written permission of both parties. Subcontracting of program research must be acceptable to both parties.

Article XV: "Notice" requires that reports or notices as required in the agreement be sent to specified individuals and addresses as designated by the parties.

Article XVI: "General Provisions" concludes the agreement. Section 16.1 prohibits the parties from using each others' names in advertising, promotional literature, news or press releases without prior written consent.

Section 16.2 states that the laws of the State of Missouri will be applied in construing the contract.

Exhibit A: "Agreement of Program Participants" represents the form which research participants are required to sign to insure compliance with relevant University policies. Subjects covered include prompt disclosure of patentable inventions, agreement not to disclose or distribute research products to others, the publications approval procedure, the necessity to avoid potential conflicts of obligations or interests by participants.

This agreement can be classified as an institution to institution agreement. It can, however, be viewed as somewhat restrictive on the University. Funds for research must be allocated with no more than thirty percent for basic research and seventy percent for projects of potential commercial utility. It could be argued that Monsanto has complete control over the selection of research topics since the granting committee has an equality of membership: four members are university scientists and four are Monsanto scientists. If Monsanto scientists did not approve a proposal it seems clear that it would not be funded since the University has only equal voting power and no clear majority.

One strength of the contract rests in the establishment of an external peer review board that meets every two years to assess the quality of the research as well as the effect of the agreement on both institutions. Financial benefits permit the University to hold all patents to marketable investors with Monsanto receiving exclusive licensing rights. Royalties are divided within the University among the Medical School, the Department, and the Laboratory. No individual researcher is permitted to profit from the agreement.

In terms of conflicts of interest issues, researchers who apply for funding under the agreement must disclose any agreements they have with other for-profit corporations. Provisions permit the sponsor to seek personal statements of confidentiality from University researchers. As for academic freedoms, the contract also has a secrecy clause stating that technical developments may not be published without the agreement of Monsanto, and, until publication, such developments remain secret for a period of two years with a two-year renewal option. Restrictions on freedom to publish permit the sponsor to review the manuscript first and require a thirty day delay of publication permitting the company to make patent application. The agreement permits Monsanto to send unlimited members of scientists to the University's laboratories. The corporate sponsor typically indemnifies the University for all licenses it receives thus protecting the University from liability.

This agreement between Washington University and the Monsanto Company is deemed significant by Barbara J. Culliton because of two characteristics:

> First, the Washington University contract is an "institution-to-institution" agreement, quite deliberately drafted to deviate from the majority of arrangements in which corporate funds are ear-marked for research by one or two senior investigators of the company's choosing.[30]

The use of the Advisory Committee composed of scientists from both institutions should avoid any divisiveness which could result if one or two prominent individual researchers controlled the research funds. Again Culliton notes:

> The second feature of the Washington University–Monsanto arrangement that sets it somewhat apart is the extent of constant, ultimate collaboration it anticipates between researchers at the two institutions. Whereas most of the new contracts contain provisions for some training of corporate scientists and for occasional interaction, this deal provides for what Howard A. Schneiderman, senior vice president of Monsanto, terms a "true partnership." Dozens of company scientists may be working on campus at any one time, once the agreement is in full swing, he notes, adding that Monsanto researchers will not be "token" members of the collaborative team.[31]

COMPARATIVE ANALYSIS

The following chart represents a comparative summary of significant common research elements, legal issues, and financial arrangements in the

Hoechst, Mallinckrodt, and Monsanto contracts which are important to institutions of higher education.

THE WASHINGTON UNIVERSITY–ANHEUSER-BUSCH COMPANIES, INC., MICROMIXING RESEARCH AGREEMENT OF 1983

Partially as a result of the successful Washington University research agreements with Mallinckrodt and Monsanto, other industries in the Saint Louis area began to seek out research affiliations with the University. Additionally, industry often desires the participation of a valued researcher who carries a faculty appointment at the University. As an example of such a spin-off agreement, examination of the Anheuser-Busch agreement of 1983 will serve as an illustration. There were actually two agreements with Anheuser-Busch, one for micromixing and another virtually identical contract for high gradient magnetic separation. The micromixing contract pays $400,000 and the magnetic separation study pays $200,000 for a combined worth in excess of half a million dollars. Since the contractual terms are similar, only the micromixing contract will be examined.

Under the terms of the agreement, Washington University laboratories will conduct basic fermentation research for the brewery. As indicated above, the combined research project will be funded at a level of $600,000.These funds are granted to the University's new Center for Biotechnology, the major research arm of the university which will coordinate the research plan. It is interesting to note, however, that one particular research supervisor is specified in both contracts—Eric H. Dunlop, Ph.D. The Washington University student newspaper carried a front-page feature story discussing the agreement in detail:

> "The Busch agreement is itself an 'enhancement' of WU's earlier accords with Monsanto Co. and Mallinckrodt Inc.," said Harry Leahey, director of WU's Research and Licensing Administration. . . .
>
> Leahey said the university's two goals for research agreements are met by the Busch pact. One goal is control over research. "We're not doing any directed research," Leahey said. "We will abide by traditional academic prerogatives," he added.[32]

A major reason for Anheuser-Busch's interest in entering into this research agreement was the reputation of the primary researcher, Eric H. Dunlop, a professor of chemical engineering.

> Dunlop is in charge of conducting the Biotechnology Center's research for Busch. Dunlop's presence at WU is one of the reasons why Busch wanted an agreement with the university, according to Randall Mayer of Fleishman-Hilliard, Inc., Busch's public relations consulting firm.

Comparative Summary of Contractual Provisions

CONTRACTUAL PROVISIONS, TERMS AND CLAUSES	Hoechst	Mallinckrodt	Monsanto
1. NUMBER OF PROJECTS	multiple but specific	multiple but specific	multiple and to be specifically identified in the future
2. ADMINISTRATIVE STRUCTURE OF PROJECTS	Scientific Advisory Board of Department; membership consists of two members from Massachusetts General Hospital; two from Hoechst and two independent	Advisory Committee; membership consists of four university members and one Mallinckrodt member	Advisory Committee; membership consists of four Monsanto members and four University members
3. TIME PERIOD OF AGREEMENT	April 14, 1981, through September 30, 1990	September 1, 1981, through August 31, 1984	July 1, 1982, through June 30, 1987
4. PROVISION FOR CONTINUATION OF AGREEMENT	automatic extension in five years unless written notice given	extended in three years by negotiated agreement	extension by mutual agreement at three ye[illegible] anniversary
5. EXTENSION OF PROJECTS IF CONTRACTS ARE NOT CONTINUED	no statement	negotiable at Mallinckrodt's request for one year	negotiable at Monsant[illegible] request for one year
6. METHOD OF SELECTION AND APPROVAL OF PROJECTS	selected by each Senior Investigator and approved by Director	selected and approved by majority of members of the Advisory Committee	selected and approved by five out of eight members of the Advisory Committee

Comparative Summary, *continued*

CONTRACTUAL PROVISIONS, TERMS AND CLAUSES	Hoechst	Mallinckrodt	Monsanto
7. RATIO OF BASIC RESEARCH TO APPLIED RESEARCH	no statement	no statement	30% basic/70% applied
8. PROVISIONS FOR INTERACTION BETWEEN SCIENTISTS	sponsor may send up to four representatives per year	no statement	sponsor may send unspecified number of representatives plus have access to university office space
9. INDEPENDENT REVIEW OF RESEARCH PROJECTS	presented for comment at annual symposium	no statement	review permitted after two years
10. FORMULA FOR ADJUSTMENT OF CORPORATE FUNDING FOR INFLATION OR DEFLATION	1981 dollars adjusted by formula—see §2.2 C	no statement	GNP Defeator Index used as part of formula—see §8.2(a)
11. TITLE TO EQUIPMENT PURCHASED WITH SPONSOR'S FUNDS	title to MGH but for Department use only	title to university	title to university at time of purchase
2. PUBLICATION RESTRICTIONS ON UNIVERSITY RESEARCHER	review by sponsor required but no control over manuscript	review by Advisory Committee with sponsor allowed to request a three month delay	review by sponsor for patentability at least one month prior to publication; sponsor may request delay of submission for a reasonable time to establish patent rights

Comparative Summary, *continued*

CONTRACTUAL PROVISIONS, TERMS AND CLAUSES	Hoechst	Mallinckrodt	Monsanto
13. OWNERSHIP OF PATENTS AND TECHNICAL DEVELOPMENTS	MGH with assignment to sponsor if MGH has no interest in title	to the university with exclusive license to sponsor	to the university with exclusive license to sponsor
14. PROTECTION PROVIDED FOR UNIVERSITY PROPRIETARY INFORMATION	no statement	sponsor to take all reasonable precautions for ten years	sponsor to take reasonable precautions
15. PROTECTION PROVIDED FOR SPONSOR'S PROPRIETARY INFORMATION	no statement	personal commitments to be made by researchers	personal commitments of confidentiality by individual participants
16. OBLIGATION OF SPONSOR TO COMMERCIALIZE RESEARCH RESULTS	if not commercialized by sponsor, license reverts to non-exclusive status	no statement	no statement
17. SPONSOR TO MONITOR RESEARCH FOR PATENTABLE AND NOVEL INVENTIONS	yes	yes	yes
18. PROMPT REPORTING AND FILING FOR PATENTABILITY BY SPONSOR	yes	yes	yes
19. PROMPT REPORTING OF POSSIBLE RESULTS BY UNIVERSITY TO SPONSOR	yes	yes	yes

Comparative Summary, *continued*

CONTRACTUAL PROVISIONS, TERMS AND CLAUSES	Hoechst	Mallinckrodt	Monsanto
20. WHO FILES FOR PATENT	sponsor	sponsor	sponsor
21. WHO FILES FOR FOREIGN PATENTS	sponsor	sponsor	sponsor
22. WHO PAYS COST OF PATENT FILINGS AND PROSECUTION	sponsor	sponsor	sponsor
23. IS PROSECUTION BEYOND PATENT OFFICE REJECTION REQUIRED	no statement	no	no
24. MAY UNIVERSITY UTILIZE INDEPENDENT PATENT COUNSEL	no statement	yes, for review only	yes, for review only
25. ROYALTY TO UNIVERSITY ADJUSTED BASED ON CONTRIBUTION OF SPONSOR	yes	no statement	yes
26. UNIVERSITY TO PROVIDE RECORDS FOR PATENT APPLICATION	yes	yes	yes

Comparative Summary, *continued*

CONTRACTUAL PROVISIONS, TERMS AND CLAUSES	Hoechst	Mallinckrodt	Monsanto
27. UNIVERSITY TO ASSURE TITLE TO ALL TECHNICAL DEVELOPMENTS	yes	yes	yes
28. UNIVERSITY WAIVER OF PATENT CLAIM AGAINST SPONSOR	no statement	no statement	yes
29. INDIVIDUAL INVENTOR WAIVER OF PATENT CLAIM AGAINST SPONSOR	no statement	no statement	yes
30. INDEMNIFICATION CLAUSE FOR CLAIMS ARISING FROM PATENT CLAIMS	no statement	no statement	yes; patent filed and prosecuted by Compan on University behalf
31. UNIVERSITY AGREES TO GRANT LICENSES TO SPONSOR	yes	sponsor has right of first refusal to license	sponsor must elect to accept license within two years of filing
32. EXCLUSIVE LICENSE ON PATENTABLE INVENTIONS TO SPONSOR	yes, if not on collaborative research	yes	yes
33. LICENSES ON NON-PATENTABLE TECHNICAL DEVELOPMENTS TO SPONSOR	no statement	exclusive for ten years	non-exclusive and non-revocable

Comparative Summary, *continued*

CONTRACTUAL PROVISIONS, TERMS AND CLAUSES	Hoechst	Mallinckrodt	Monsanto
34. LICENSING OF NON-PROGRAM PATENTS BY UNIVERSITY TO SPONSOR	right to best possible license	no statement	yes, to best extent possible
35. LICENSING REQUIREMENTS SPECIFIED IN AGREEMENT	yes	yes	yes
36. REASONABLE ATTEMPT TO BE MADE BY SPONSOR TO MARKET RESEARCH RESULTS	indirectly	yes	yes
37. IF REASONABLE EFFORT NOT MADE BY SPONSOR TO MARKET RESULTS, THE NON-EXCLUSIVE SUBLICENSE RESULTS	yes, after three years	no statement	yes
38. INDUSTRY SPONSOR REQUIRED TO SUBMIT MARKETABILITY SCHEDULE DURING PERIOD OF EXCLUSIVE LICENSE	no statement	yes	yes
39. SPONSOR PERMITTED TO SUBLICENSE	no statement	yes	yes, with notification by university
40. ROYALTY PAYMENTS BY SPONSOR TO UNIVERSITY	not to exceed 50% of the fair commercial royalty rate for such license (6.4)	royalty schedule based on Net Selling Price with a rate scale (8.14d)	royalty rate to be negotiated at licensing

Comparative Summary, *continued*

CONTRACTUAL PROVISIONS, TERMS AND CLAUSES	Hoechst	Mallinckrodt	Monsanto
41. IF NO ROYALTY RATE AGREEMENT, ARBITRATION IS PROVIDED	yes	no statement	yes
42. LAW TO BE APPLIED TO AGREEMENT	Massachusetts	Missouri	Missouri
43. ACTION TO BE TAKEN IN EVENT OF INFRINGEMENT	no statement	no statement	sponsor to obtain discontinuance or bring suit (11.18.1(1))
44. WHO MAY SUE INFRINGER?	no statement	no statement	Sponsor may sue in its name or university's name; university may s[illegible] in its name if sponsor does not sue
45. INFRINGEMENT SUIT COST RECOVERY	no statement	no statement	Settlement or recove[illegible] shall first be used to [illegible] expenses of the party bringing the suit; bal[illegible] to be distributed 2/3 party bringing suit an[illegible] 1/3 to other party.
46. HAS UNIVERSITY RIGHT OF APPROVAL PRIOR TO SPONSOR BRINGING SUIT?	no statement	no statement	yes
47. UNIVERSITY MAY ASSIGN TITLE TO SPONSOR BRINGING SUIT	no statement	no statement	yes

Comparative Summary, *continued*

CONTRACTUAL PROVISIONS, TERMS AND CLAUSES	Hoechst	Mallinckrodt	Monsanto
48. RIGHT OF UNIVERSITY TO LICENSE ELSEWHERE IF SPONSOR DOES NOT ELECT TO LICENSE	no statement	yes, with sponsor receiving 50% of royalties	yes, but no license permitted to sponsor's competitors
49. TERMINATION OF AGREEMENT FOR BREACH OR DEFAULT	no statement	90 day written notice unless breach is cured	90 day written notice
50. TERMINATION FOR INSOLVENCY OF A PARTY	no statement	30 day written notice	30 day written notice
51. PATENT AND LICENSE RIGHTS SURVIVE TERMINATION OF AGREEMENT	no statement	yes	yes
52. INDEMNIFICATION OF UNIVERSITY BY SPONSOR FOR LIABILITY ARISING FROM USE OF PRODUCTS	no statement	yes	yes
53. UNIVERSITY WARRANTS SUFFICIENT INSURANCE AND WORKMEN'S COMPEN- COMPENSATION FOR ITS EMPLOYEES	no statement	no statement	yes
54. SPONSOR TO HOLD UNIVERSITY HARMLESS FOR SPONSOR'S EMPLOYEES INJURED AT UNIVERSITY	no statement	no statement	yes, if no negligence by university

Comparative Summary, *continued*

CONTRACTUAL PROVISIONS, TERMS AND CLAUSES	Hoechst	Mallinckrodt	Monsanto
55. ASSIGNMENT OF RIGHTS AND OBLIGATIONS BY EITHER PARTY	yes, if to a corporation controlled by one of the parties and with a guarantee from the party of the performance of its assignee	no	no
56. MAY UNIVERSITY SUBCONTRACT	no statement	no statement	only with approval of sponsor
57. USE OF UNIVERSITY OR SPONSOR'S NAME WITH PUBLICITY	no statement	may use other party's name only with permission	use of other party's na[me] only with permission
58. RESEARCH PARTICIPANT NON-DISCLOSURE AGREEMENT	yes required	yes	yes
59. DISTRIBUTION OF PRODUCT BY UNIVERSITY FOR RESEARCH ONLY	no statement	yes	no statement
60. THIRD PARTY COMPETITOR DISTRIBUTES SIMILAR PRODUCT SIGNIFICANTLY AFFECTING SPONSOR'S BUSINESS, THEN ROYALTY PAYMENT TO UNIVERSITY MAY CEASE	no statement	yes	no statement
61. REPAYMENT OF SPONSOR'S COSTS REQUIRED AT TERMINATION OF AGREEMENT	fair market value of equipment and furniture plus half of cost of construction of building to be repaid	no	no

Comparative Summary, *continued*

CONTRACTUAL PROVISIONS, TERMS AND CLAUSES	Hoechst	Mallinckrodt	Monsanto
62. PROVISION FOR ALTERNATIVE FUNDING TO UNIVERSITY	sponsor has exclusive right to fund research unless sponsor declines	no statement	no statement
63. CONTRACT FUNDING STATED	$67.3 million over 10 years	$3,881,250 over three years	$23.5 million over five years
64. DESIGNATED DIRECTOR OF RESEARCH OR PRINCIPAL INVESTIGATOR	MGH to appoint with consultation of sponsor; Dr. H. Goodman initially designated	Dr. Joseph M. Davie	appointed by university—Dr. D. Kipnis
65. PROVISION FOR RESOLUTION OF DISPUTES	none; Joint Committee of three trustee members of MGH and three senior executives of Hoechst serve to oversee agreement and forum for communicaton	no statement	Royalty rate disputes may be submitted to binding arbitration under rules of American Arbitration Association

Mayer cited as the brewery's primary motivation "the national and international reputation of Eric Dunlop in biotechnology."

Mayer added that WU was chosen by Busch because of "the quality of the school." Location in St. Louis was also a consideration, he noted. "Closeness will facilitate cooperation" and sharing of information, he said.[33]

The agreement consists of twelve pages without attachments or appendices. The following abstract and analysis represents an article-by-article summary of the significant sections of this contract.

The general introductory recital establishes the desire of the Anheuser-Busch Companies, Inc., to participate in research in the area of biotechnology and biochemical engineering as it relates to fermentation and to apply these research results to its business. The recital notes the desire of Washington University's Center for Biotechnology under the leadership of Eric H. Dunlop to undertake biotechnological research and development in the area of "micromixing." The element of consideration creating a binding contract is identified as the mutual promises made within the contract.

Article 1 defines "micromixing" as encompassing advanced biochemical engineering technology as it relates to the mixing of fermentation substances on a micro-scale, with a goal of developing more effective ways to provide nutrients and to remove metabolic waste from yeast.

Article 2 is entitled "Research Subjects" and notes the goal of the agreement as improving and facilitating the mixing of fermentation substances in production and operational situations. Specifically, the objective of the research is to study ways of implementing micromixing techniques in the brewing industry so as to provide practical improvements and economic benefits to the production process. This section specifies that the research leadership is to be provided by Dunlop and that he has been selected because of his particular expertise and academic credentials. Persons other than Dunlop involved in the research such as students, research scientists, and technicians are to operate under his direction and supervision. As part of the agreement the University agrees to require Dunlop to devote not less than fifteen percent of his annual time to the research project.

Article 3 deals with "Financial Terms." This agreement specifies its limits at $400,000 payable in four yearly installments of $100,000, beginning with the date of the contract's execution. The funds are controlled by the Center for Biotechnology and are for the purchase of supplies, to defray expenses for scientific meetings, and to support graduate students, research scientists, and technicians.

Article 4 addresses the "Term" of the contract. The contract becomes effective on the date of execution and terminates on the fourth anniversary date. Renewal will be evaluated at the conclusion of the term of the contract with the desire that practical application will be achieved during the four year period.

Article 5 establishes a "Joint Advisory Committee." The Committee's function concentrates on coordinating and interfacing the joint research and development efforts in a spirit of mutual trust. The Committee will provide a mutual assessment of the research being conducted, seek to resolve problems, and assist with scientific interfacing. Membership on the Committee consists of three university representatives with Dunlop specifically identified as chairman. Three additional representatives are appointed by Anheuser-Busch. The Committee will meet at least once every three months and review allocations of the grant fund. The Committee may recommend that funding be expanded beyond its financial restrictions or it may recommend redirection of funding to another project, if warranted. In all instances, and unlike other industry–university contracts, the final decision regarding the Committee's financial recommendations rests solely with Anheuser-Busch.

Article 6 requires *Progress Reports*. Reports are required at least quarterly to coincide with the meetings of the Joint Advisory Committee. In addition, an annual Progress Report is required in June of each year and shall include a comparison of monies granted with monies spent and allocated.

Article 7 specifies ownership and use of *Equipment*. The University agrees to make available its equipment and materials through the Center for Biotechnology to assure a successful research endeavor. Anheuser-Busch will grant university researchers access to company equipment and materials, provided there is no interference with corporate production activities.

Article 8 addresses *Confidentiality* of information. The first three paragraphs state the understanding that Anheuser-Busch recognizes that a variety of information which comes into its possession through the agreement constitutes valuable university property and will exercise the same safeguards it uses to protect its own confidential technical information. The parties agree that, despite precautions, should an unauthorized release of information result, Anheuser-Busch will not incur liability. The final two paragraphs reference the University's commitment to protecting corporate confidential information by advising all project participants that they will be required to sign personal, confidential non-disclosure agreements.

Article 9 states a *No Conflict* agreement in which the University participants agree not to accept consultation or grant assignments from any other party involved in micromixing related to brewing, wine making, baking or production of baker's yeast including recombinant DNA yeast and genetically modified yeasts. The University states that this agreement does not conflict with existing research agreements. By this Article, final decision as to a conflicts question rests solely with Anheuser-Busch.

Article 10 articulates the parties' rights to *Patents and Licenses*. The University will pursue Letters Patent and arrange for preparation, filing, and prosecution of patent applications. Anheuser-Busch will reimburse the University for reasonable costs incurred in patent searches. The corporation reserves the right to give approval and instruction to attorneys or to the Patent and Trademark office in situations where it might incur a reimbursable expense. The University grants the corporation an exclusive royalty-free license in perpetuity with the right to sublicense in consideration of the corporation's covering patent costs and expenses. If after ten years of receiving the initial disclosure Anheuser-Busch has not incorporated the product or process into its commercial operations, then the license becomes nonexclusive. If the corporation sublicenses the invention to a third party, the University shall receive a fifty percent royalty.

Article 11 states restrictions on *Publication*. Recognizing that scientific discoveries made in the cause of research may be freely reported, Anheuser-Busch reserves a thirty-day period in which to review an abstract of any proposed report. The corporation has the right to delay publication for a reasonable time in order to protect its intellectual property rights. Following the filing of a patent application claiming the discovery in question, publication becomes permissible.

Article 12 sets forth methods of adjusting the agreement for *Significant Changes*. These changes look specifically to the problems which might develop due to the absence of Dr. Dunlop as chief overseer and expert.

If Dunlop is absent in excess of two months the Joint Committee will make recommendations as to the disposition of the research project. All decisions on continuance and funding are reserved solely with Anheuser-Busch.

Article 13 addresses *Indemnification*. Both parties will hold each other harmless and indemnify the other from losses, damage, and liability arising from negligence on the part of either party's employees. Anheuser-Busch will indemnify the University from liability and losses from the use or sale of any product developed from or based upon information or materials received from the University within the scope of the contract.

Article 14 covers developments beyond the control of the parties under the title *Force Majeure*. Should circumstances beyond the control of the parties prevent the performance of research as agreed, both parties will make every effort within their control to rectify the situation and to fulfill their obligations.

Article 15 states procedures for *Early Termination and Renewal* of the agreement. Anheuser-Busch reserves the option to terminate this agreement should the University or Dunlop discontinue to conduct research which forms the basis for the contract. Before termination, the parties agree to discuss mutually through the Joint Advisory Committee any apparent discontinuance on the part of the University. In any event, either party may terminate the agreement by a written twelve-months advance notice.

One unique aspect of this agreement is an automatic renewal clause. Anheuser-Busch agrees to grant the University an additional $400,000 grant if its net profits exceed $400,000 within ten years of disclosure as a result of the commercial application of discoveries from the research produced through this agreement.

The final Articles are general in nature: Article 16 states that *Governing Laws* for the contract will be those of the State of Missouri. Article 17, entitled *Notices*, specifies addresses of contract persons for written notices. Article 18 covers *Assignments* stating that obligations are not assignable to other parties and subcontracting by the University or Dunlop shall not be permitted.

Article 19 is an *Entire Agreement* clause stating that all promises are contained in the contract. Article 20, *Amendments*, permits modification by a written instrument signed by authorized representatives of all parties. Article 21, *Headings*, indicates article headings are for convenience and do not limit the scope of the paragraphs to which they pertain. The contract contains signatures of Washington University's Associate Vice Chancellor Edward MacCordy, Professor Eric H. Dunlop, and John H. Purnell, Vice President of Anheuser-Busch Companies, Inc., and was dated 15 August 1983.

In comparison to previous contracts this agreement is clearly institutional-to-individual. Corporate control is very clear. It does not appear weighed in favor of the University. The corporate sponsor specifically earmarks its funds for research by one senior investigator and acknowledges that it was induced because of Professor Dunlop's expertise and academic credentials. The agreement specifies that Dunlop will directly supervise and be involved in the research. The University is required to permit Dunlop to spend no less than fifteen percent of his annual time on this one research project. A Joint Advisory Board oversees the research. Its membership consists of six individuals—three representatives from the University including Dunlop as chairman and three representatives from the Anheuser-Busch Companies. The Board's roll is strictly advisory with the agreement stating that all final decisions will be made by Anheuser-Busch Companies. Confidentiality of corporate data and proprietary information is protected by personal commitments of confidentiality and nondisclosure which the sponsor may require as it deems necessary. These statements of confidentiality are common practice in industry but very unusual in the university setting where they may conflict with academic freedom concepts.

Questions of conflicts of interest are firmly in the sponsor's hands. The agreement places final decisions as to whether or not a conflict exists with a specific research interest with Anheuser-Busch Corporation. No conflicting general research is permitted by Dunlop and no consulting assignments may be undertaken without the written consent of the sponsor. Unlike other agreements, the University and not the sponsor will seek patents. The sponsor, however, will have an exclusive royalty-free license in perpetuity with rights to sublicense in consideration of its paying for Dunlop's research. Interestingly, this agreement is subject to automatic renewal if the sponsor's net profits surpass its initial $400,000 investment. Publication of research is permitted after an abstract is approved by Anheuser-Busch. The sponsor has thirty days in which to issue its approval or ask for a delay in publication for a period specified only as reasonable time.

As an example of a smaller spin-off contract from the larger agreements into which Washington University has entered, this agreement has some definite limitations. Harry Leahey, Director of Washington University's Research and Licensing Administration, defines it as "not an optimal agreement":

> In terms of royalties, this agreement does not place the university in an optimum position. It involves one specific project supervisor. The final decision on financial recommendations, conflicts questions, and significant changes in the contract rests solely with executives at Anheuser-Busch and not with the Joint Advisory Board or with the corporation's own Research Department. One unique enhancement, however, is that the agreement becomes automatically renewable for the full amount if Anheuser-Busch profits by the research results beyond its $400,000 initial investment. We traded royalties for an automatic reinstatement of the agreement at full value should our technology reduce costs or increase profits by a certain percentage. Believe me, we will take a renewal of an agreement every time over

royalties—royalties have never been a significant source of income—whereas the $600,000 renewal of the agreements would be.[34]

OVERVIEW

The ultimate success of the industry–university contracting approach can be measured by the eventual products that reach the market place. Such research that results in commercially successful products should not detract or distort the university's academic goals of teaching, research, and service. Samuel B. Guze, Vice Chancellor for Medical Affairs at Washington University, establishes the following guidelines:

> Such distortion can be minimized if everyone is informed, if such research represents only a small portion of the university's total research effort, if participation in such research is voluntary, if the quality of such research is high, as judged by disinterested scientific peers, and if there is no personal financial gain to the university personnel involved in the research.[35]

It is through the use of contracts of agreement that institutions of higher education and industry can establish their mutual research goals while protecting all parties' interests. The final benefactor of these relationships as created in contracts will be society.

Dr. Schneiderman of Monsanto stated the ultimate result as follows:

> By combining two of America's greatest resources—the scientific expertise of one of the country's leading medical schools and the technology and scientific capability of a large high technology corporation—we have the opportunity to create new business and new jobs and advance technology. Through this collaboration we expect to discover novel products which address major human disease and health conditions for which there is presently no cure and no adequate therapy. We view this aim as both socially responsible and commercially attractive.[36]

These contracts for cooperative research ventures are still, however, in the experimental stage. The desire for success exists on the part of the participants. In the final analysis, the true test of the acceptability of these research relationships will be measured by the number of agreements that are renewed or expanded. This, in turn, will be contingent upon whether industry's investment in university research provides a financial return.

NOTES

1. See Rustum Roy, "University–Industry Interaction Patterns," *Science* 178 (1 December 1972):955–960.
2. Wayne Biddle, "A Patent on Knowledge: Harvard Goes Public," *Harper's*, July 1981, pp. 22–26.

3. Allen L. Hammond, "Methanol at MIT: Industry Influence Charged in Project Cancellation," *Science* 190 (21 November 1975):761–764. See also, letters to the editor, *Science* 191 (30 January 1976):338–339.

4. Derek C. Bok, "Universities and National Research Policy," *Science* 193 (10 September 1976):955.

5. R. Jeffrey Smith, "Patent Policy Changes Show Concern," *Science* 199 (17 March 1978):1190.

6. Marjorie Sun, "NIH Ponders Pitfalls of Industrial Support," *Science* 213 (3 July 1981):113–114; Barbara J. Culliton, "The Hoechst Department at Mass. General," *Science* 216 (11 June 1982):1200–1203.

7. Barbara J. Culliton, "Pajaro Dunes: The Search for Consensus," *Science* 216 (9 April 1982):155.

8. Thomas W. Langfitt et al., eds., *Partners in the Research Enterprise: University-Corporate Relations in Science and Technology* (Philadelphia: University of Pennsylvania Press, 1983).

9. Barbara J. Culliton, "The Academic-Industrial Complex," *Science* 216 (28 May 1982):960–962.

10. Edward H. Bowman, *University Investing and Corporate Responsibility* (Cambridge, Mass.: Massachusetts Institute of Technology, 21 August 1971), p. 36.

11. Marjorie Sun, "NIH Ponders Pitfalls of Industrial Support," p. 113.

12. Robert D. Varrin and Diane S. Kukich, "Guidelines for Industry Sponsored Research at Universities," *Science* 277 (25 January 1985):385–388.

13. Charles C. Caldart, "Industry Investment in University Research," *Science, Technology, & Human Values* 8 (Spring 1983):27.

14. Ibid., p. 28.

15. Preston W. Grounds, *University–Industry Interaction: Guide to Developing Fundamental Research Agreements* (Allentown, Pa.: Council for Chemical Research, 1983), pp. 6–9. See also Milton Strombler, *Guide to University-Industry Research Agreements* (Santa Monica, Calif.: Society of Research Administrators, 1983).

16. Caldart, "Industry Investment in University Research," p. 28.

17. Barbara J. Culliton, "The Hoechst Department at Mass. General," p. 1200.

18. Ibid., p. 1201.

19. Ibid., p. 1202.

20. U.S. Congress, House, Committee on Science and Technology, *Commercialization of Academic Biomedical Research, Hearings before a Subcommittee on Science, Research, and Technology*, 97th Cong., 1st sess., 1981, pp. 87–88.

21. M. J. Socolar, Comptroller General of the United States, letter addressed to the Honorable Albert Gore, Jr. (16 October 1981), ref. B–204687.

22. See "Allies in the Laboratory: University and Industry Scientists Create a Model for Research Agreements," *Washington University Magazine* 53 (Summer 1983):19–22.

23. Kathy Liszewski, "A New Weapon in War on Disease," *St. Louis Post-Dispatch*, 1 June 1982, p. D1.

24. Dale Singer, "Can Industry Control Campus? The Academic-Industrial Complex," *St. Louis Post-Dispatch*, 26 September 1982, p. A13.

25. Ibid.

26. Dale Singer, "Biomedical Research 'Marriage' Here," *St. Louis Post-Dispatch*, 28 September 1982, p. Al. See "Allies in the Laboratory: University and Industry Scientists Create a Model for Research Agreements," *Washington University Magazine* 53 (Summer 1983):19 ("The result is an agreement which . . . may serve as a prototype for other university/industry ventures of this kind.").

27. U.S. Congress, House, Committee on Science and Technology, *University/Industry Cooperation in Biotechnology, Hearing before the Subcommittee on Investigations and Oversights,* 97th Cong., 2d sess., 1982, p. 12.

28. Ibid., p. 21.

29. "Allies in the Laboratory," p. 22.

30. Barbara J. Culliton, "Monsanto Gives Washington U. $23.5 Million," *Science* 216 (18 June 1982):1295.

31. Ibid.

32. Louie Carlot, "Busch Funding Sparks Biotechnology Research," *Student Life*, 27 September 1983, p. 1.

33. Ibid.

34. Interview with Harry Leahey, Washington University, St. Louis, Missouri (29 September 1983).

35. Samuel B. Guze, "The Monsanto–Washington University Biomedical Research Agreement," in *Partners in the Research Enterprise: University-Corporate Relations in Science and Technology,* ed. Thomas W. Langfitt et al. (Philadelphia: University of Pennsylvania Press, 1983), p. 57.

36. *University/Industry Cooperation in Biotechnology, Hearing before the Subcommittee on Investigations and Oversights*, p. 21.

5

Future Trends with Industry–University Innovation Agreements

THE FUTURE POTENTIAL FOR CONTRACTING

Corporate funding of university research in its present guise and involving multimillion dollar investments is a relatively new development. The preceding contracts clarify the kinds of concessions that universities will make in order to receive corporate funds. These agreements attempt to specify methods of protecting the university's traditional research freedoms while assuring industry that its proprietary interests will be safeguarded. Uncertainty still exists as the world watches the current agreements in operation. This lack of clarity is marked by an ambivalent attitude concerning those who recognize that corporate sponsored research offers timely advantages. In the Congressional hearings on University–Industry Cooperation, Representative Walgren (D. Pa.) captured the mood of skepticism that still surrounds corporate sponsored research:

> [I]mproved relationships between universities and industries could help to foster innovation and increase productivity in the U.S. . . . However, in recent years we have been faced with a dilemma. Many of the developments in molecular biology, genetics, and cell biology can be commercialized almost immediately. . . . This research in the biotechnologies has the potential to generate funds for universities, many of which are actively seeking alternative funding sources. Yet, the very process of developing this potential may change the fundamental mission of universities, that is, education and research in an atmosphere of freedom of thought and free exchange of ideas.[1]

Government officials and academics alike are concerned about the contamination that competition produces in the industry and its possible transfer to the university environment. But competition can also enrich the academic community. And collaboration and cooperation can do much to open the frontiers of knowledge. Joint research efforts in innovation can, if properly articulated, accomplish goals of mutual interest to both industry and the university—the improvement of society.

Barbara Culliton, news editor of *Science* magazine, in comparing the Hoechst and Monsanto agreements made the following observations:

First, agreements of any substantial dollar amount are few in number. Second, those that do exist . . . are different from each other. Third, they are clearly experimental in nature, and the contracting parties recognize that fact. Finally, although they are written in each instance by people on both sides dedicated to the proposition that academic freedom and openness of communication would be maintained, we do not yet know whether that will prove to be true in the end.[2]

Culliton stresses the experimental aspects of these two agreements and difficulties that are clear on the face of contracts. Two factors appear to be mutually exclusive. She observes that the Hoechst agreement tries to protect the rights of communication between researchers and assure the open publication of research results. On the other hand, heavy emphasis is placed on the nature of the contract as an exclusive agreement and the retention by the company of exclusive patent rights to research results.[3] In comparing the Monsanto agreement with the Hoechst contract another potential problem exists:

In reading the Hoechst agreement . . . I note that the company has the right to have four of its scientists at the hospital at any given time. With that in mind, I asked the Monsanto–Washington University people how many Monsanto scientists they anticipated might be physically working on the university campus at any one time. Well, said the man to whom I was talking, I'm not sure. I asked, four or five? No, more than four or five. A dozen? More than a dozen. Two dozen? Likely more than two dozen.

Therefore, one of the most important parts of the relationship to watch is the effect of the influx of a large number of industrial scientists into academic laboratories.[4]

In addition to the betterment of society, industry is interested in academic research because it is good business. There is a legitimate self-interest by corporations which must be recognized. The existing contracts by their terms are highly specialized and specific. As has been stated, the final test of their usefulness to industry will be their continued use.

The scope of this discussion has been the bilateral research relationships between the university and the company or an individual scientist and the industry. Multi-institutional relationships involving a number of universities and corporations and occasionally the federal government have not been explored. A thorough study of government involvement in the research and development process and the significance of industrial innovation policy from a historical case method approach has previously been examined by Professors Richard R. Nelson and Richard N. Langlois.[5] Readers desiring additional information on the federal government's role in the innovation process should refer to that essay.

MODELS, RULES, AND GUIDELINES: TRENDS IN CONTRACTING OF INNOVATION

Innovation is the ultimate process which joint industry–university agreements are seeking to accomplish. Innovation has been defined by researchers as follows:

> Innovations are the units of technological change. A technical innovation is a complex activity which proceeds from the conception of a new idea to a solution of the problem and then to the actual utilization of a new item of economic or social value. . . . Innovation is not a single action but a total process of interrelated subprocesses.[6]

A variety of other models of research agreements of the bilateral type has been identified. Thomas W. Langfitt, President for Health Affairs at the University of Pennsylvania, listed representative examples which are identified as follows:[7]

1. Consulting:[8]
   The company employs a faculty member to consult with the company during a portion of the time that the university allows the faculty member to work outside the university. Ordinarily the faculty member keeps all income paid and does not report it to the university. Washington University Technological Associates is an example of a university association created to develop and market consulting skills of faculty.
2. Research Contract between the University and the Corporation:[9]
   A corporation purchases a portion of a university scientist's time and uses his team to help do applied research in either the university or the industrial laboratory. The university benefits from royalties gained in the sale of any product that emanates from the relationship.
3. Large University-Corporate Agreements:
   . . . [T]hese agreements typically involve large sums of money. Some critics view these agreements as a threat to the integrity of the university's scientific efforts. Because the ultimate applications of the research are often not clearly formulated when the agreement is signed, modifications and constant monitoring of the research are expected. Due to high monetary investment by the company, restrictions on the research team may be more than normal.
4. Affiliate Programs:
   Pioneered by Massachusetts Institute of Technology, industrial scientists, for an annual fee, tap into the university laboratory for new ideas or help with a specific problem.
5. Exchange of Scientists:
   A laboratory group in the university and a corporate laboratory with similar interests agree to exchange scientists for specific periods of time. This exchange known as "job trading" has been tried at Clemson and Purdue.[10]
6. Venture Capital Companies Spawned by University Scientists:[11]
   Usually the sequence of events is as follows: the university scientist develops

a marketable idea, and the university patents it. Through personal connections, the scientist comes in contact with a venture capitalist. If the university has clearly established guidelines for licensing patents and sharing royalties, and for dealing with conflicts of interests of its employees, most of the issues that arise from this point onward can be confronted directly and properly managed. If there are not adequate guidelines, the scientist and venture capitalist may be far along with the development of a product before the university is aware of what is happening. At this point the scientist, and perhaps several members of his laboratory, have a large equity in the new company, and the university has been left out or its share is small.

A seventh example which can also enhance the Langfitt list is identified by our federal government:

7. Research and Development Limited Partnerships:
Recently advocated by the U.S. Department of Commerce, this partnership is governed by a partnership agreement. This agreement provides for two classes of parties, general and limited. The general partners provide the management for the business, obtain funding, make arrangements for the conduct of research, and ultimately manufacture new products resulting from the research, or licenses the research results. The limited partners are investors in the business but exert no active management. The limited partners provide capital in exchange for the benefit of tax shelters generated by the partnership's research and development expenditures.[12]

Multi-institutional relationships can involve the following types of agreements:

1.The Research Park: University surplus campus or endowment land is made available to industry on which to build laboratories. Industry is attracted to the park by the presence of scientists on the campus who wish to cooperate with industrial scientists. According to Larry R. Thompson, Special Assistant to the President, Ohio State University is exploring the development of such a plan.[13] The Research Triangle enterprise in North Carolina illustrates a similar arrangement.

2.The Industry–University Cooperative: This approach involves a cooperative program under the sponsorship of National Science Foundation, University–Industry Cooperative Research Centers Program. Most centers develop a varied "menu" of research topics and issues around the core technology. At each center there are usually four to twelve research projects underway at any given time.[14] Examples include the Center for Welding Research at Ohio State University and the Center for Interactive Computer Graphics at Rensselaer Polytechnic Institute.

3.The Stanford Industrial Innovation Centers: Founded by various industrial sponsors whose research monies are pooled, the center serves as an accumulation point and vehicle for university researchers to access industry. Examples include the Center for Integrated Systems being funded by nineteen major American computer and electronics companies that contribute $750,000 each for a total of $14.25 million.[15]

A second arrangement involves the funding of biotechnology research at Stanford, primarily in the Department of Chemical Engineering. It was proposed that a nonprofit organization would be formed with contributions from sponsoring companies, which organization would form a for-profit company to do product-development research with rights to the resulting technology eventually being licensed to the sponsoring companies. According to staff counsel Adrian Arima of the General Counsel's Office, for conflict of interest reasons, Stanford elected to participate in the organization and control of the nonprofit organization.[16] The nonprofit Center for Biotechnology Research eventually was formed independent of Stanford, and it in turn controls the for-profit Engenics Company. The Center funds research at Stanford (as well as other universities) under standard research agreements.

The establishment of university created rules and guidelines serves as the logical foundation on which the contractual research relationship must be firmly based. Thomas H. Moss, Director of Research Administration at Case Western Reserve University, has identified the lack of centralized rule making of industry–university research agreements by the federal government as a positive sign for university self-regulation:

> In the case of industrial innovation, despite the tendency to hope for a single strong mandate around which to rally, colleges and universities will probably be left with the need to seek individual, local, and regional solutions. The bad news is that no one is likely to step in and magically guide us—nor is the federal government likely to incorporate us into a grand strategy. The good news is that we will have every opportunity to use our strengths and creativity to seek our own solutions.[17]

Efforts at establishing these solution oriented guidelines are underway. Langfitt and others have proposed a series of rules to provide guidance to those parties who seek to establish industry–university research agreements. Most of these rules center on methods of disclosing research relationships between faculty members and corporations, publicity regarding contractual arrangements, ownership and licensing of patents, handling of corporate proprietary information, and time from a research discovery to publication in a scientific journal. Specifically, each university is encouraged to create and publicize its guidelines for corporate research relationships and establish a procedure for examination of specific agreements so as to assure their compliance with the guidelines.[18]

On 3 November 1982 the University of California System issued its "Interim Guidelines in University–Industry Relations." The interim guidelines constituted the university's policy of preserving the university's mission of teaching, research, and public service. Its goal was to maintain institutional independence and integrity to assure impartiality, to maintain an environment that permits faculty and students freely to pursue learning and research, and to assure that the trust of the public be kept.[19]

The American Civil Liberties Union issued on 28 October 1983 its own guidelines on university contract research emphasizing research relations between corporations and academic institutions.[20] The primary function of the ACLU guidelines focuses on protecting the values of academic freedom as it affects faculty and students.

Private universities such as Stanford have established a variety of guidelines to instruct their faculty. These include "Policy on Tangible Research Property," "Conflict of Interest Situations," "Policy and Procedures Summary for Sponsorship of Research," and "Policy on Consulting by Members of the Academic Council."[21]

At the national level, the Association of American Universities (A.A.U.) was requested by Representative Albert Gore, Jr., on 18 November 1981 to address the problem of ethical considerations which might arise from cooperative research interests between corporations and universities by drafting ethical guidelines. In September 1983, the Association established its Clearing House on University–Industry Relations. The Committee determined that guidelines were unnecessary but that universities, industry, Congress, and the public would benefit greatly from sharing information regarding research collaborations.

On 10 January 1984 Robert M. Rosenzweig, President of the A.A.U., sent letters to all member institutions requesting information on two topics.

> We are interested in receiving written information concerning your university's policies and practices, including documentation of policy, such as statements, guidelines, and memoranda, and discussions and documentation of practices, including contracts and other agreements. We are not requesting confidential information. . . . We hope to receive information covering the breadth and variety of university activities in this area while including the details of specific arrangements.[22]

The A.A.U. has just begun to collect data on policies and contracts. At this writing it is too early to predict the usefulness of this data and the role of the Clearing House.

REFLECTIONS ON CURRENT CONTRACTS

Thomas H. Moss, Director of Research Administration at Case Western Reserve University, forecasts that these new industry–university research partnerships would be a boon for higher education:

One key component of a uniquely American strategy for innovation is beginning to emerge. It is the enormous burst of industry-university cooperation in research and development leading to new products and technology.

The potential for making money—through the direct support of research by industry, income from royalties, or other material benefits of marketed technology—is probably the principal reason that universities are interested in such relationships.[23]

Jack Duvall, Director of Corporate Relations at the University of Chicago, has challenged this view, noting that a few significant agreements between selected prestigious research universities and a few of the largest corporations hardly qualifies as a ground swell:

But beyond recouping current costs of research, the chance of realizing appreciable profits in the future from patents obtained from work performed with corporate support is still very speculative. It would hardly be the principal reason why faculty members would agree to accept outside support.

For their part, sponsoring companies are interested mainly in having personal access to university faculty members who are conducting advanced research, not because they foresee discrete breakthroughs, that can be hustled to the marketplace, but rather because their long-term success as technological competitors depends on exposing their own scientists and engineers to the influence and ideas of first-rate university researchers. The phrase often used when these agreements are being negotiated is that the company wants to "buy a window" on work at the university.[24]

To determine the success and present status of the Hoechst, Mallinckrodt, and Monsanto agreements, personal interviews were made with representatives of the respective parties. Their reflections are summarized in the following personal statements.

Harry W. Orf, Ph.D., Deputy Director of the Department of Molecular Biology at Massachusetts General Hospital reviews the Hoechst agreement as follows:

No problems have been experienced with Hoechst. Any concern that the contract of agreement implied corporate control over research has proven untrue. Researchers have complete academic freedom. There is no corporate involvement in the day-to-day program or in the overall scientific program. There does exist an ongoing interaction in the scientific context where department scientists travel to Germany to present seminars to Hoechst scientists. These seminars often do not relate directly to Hoechst's marketing interests. As one with doubts when joining the Department, I believe academic freedom has held up extremely well. The Department has proven to be a completely independent academic unit and has no tinge of corporate control. As for the corporation's satisfaction with the agreement, I have no experience on which to base a positive or negative answer. I will comment that the money keeps coming in.[25]

David Kipnis, M.D., Director of the Biomedical Research Agreement with Monsanto commented as follows on both the Mallinckrodt and Monsanto relationships:

The renewal of the Monsanto Hybridoma agreement is in question. Monsanto's research interests are forming in other areas and hybridoma research may not be a future area in which they wish to concentrate. The Mallinckrodt Hybridoma research agreement is also in question and renewal is not assured. With Dr. William Phillips of the Washington University faculty currently serving as their director of research the Mallinckrodt agreement may be renewed, but with greater limitations, restrictions, or a different focus.

In general the institutional success of the broad biomedical research agreement with Monsanto has exceeded expectation. Exploratory areas were developed by faculty and innovative investigations undertaken. Initially, the approach used by the medical school involved an invitation which took place within the first six months of the agreement. This invitation went to medical school faculty who were prominent researchers. They were urged to submit proposals articulating areas of investigation which might fall within the scope of the Monsanto contract. A competitive option was then extended to all medical school faculty. Under this plan all faculty who are interested in making a research proposal are free to compete and submit their proposals to a review board.[26]

Dr. Kipnis notes that the Monsanto contract is uniquely different in various areas. First, this agreement does not identify specific individuals, nor was it drafted for a specific group of individuals. The Hoechst agreement identified Dr. Goodman directly and named him as the institute's director. Second, the Monsanto agreement has a built-in peer review system which is both internal and external. A third characteristic addresses the speedy publication of research results after a short, thirty-day review period. A fourth characteristic reflects the fact that the Medical School has no practice plans. No individual member of the research team benefits in the way of a salary increase or bonus. All royalties and profits generated through the research agreement are distributed as follows: 20 percent to the Dean of the Medical School, 40 percent to the department in which the major researcher resides, and 40 percent to the laboratory.

Edward L. MacCordy, Associate Vice Chancellor for Research at Washington University described the non-renewal process as follows:

Corporate product planning is subject to change. What happens when corporations shift from one area of product development for the future to another represents their prediction of future market trends. Monsanto's hybridoma research agreement will not be renewed due to the shift in their initial interest in the hybridoma market. Monsanto decided to enter the health care field with new product lines for the future. They are staffing up and acquiring facilities. Research in the new hybridoma technology initially was of potential health care research interest. But a continuing review of their long range market plans resulted in their decision that hybridoma based diagnostics is not an area for them to become involved in now. By the time products are developed they will be behind the general market trends. Thus, Monsanto will not continue research support in hybridoma technology beyond the present contract term.

Mallinckrodt also is not electing to renew their hybridoma research agreement. Mallinckrodt has indicated that for the future they will not be emphasizing clinical diagnostics and would be shifting their product interest to items sold directly to physicians' offices. Mallinckrodt states that is is deemphasizing monoclonal antibody research for the U.S. market.

As a result of the non renewal of these two contracts the university has determined that major industrial agreements must be for a reasonable term. Washington University now looks to a minimum agreement for a five year period as opposed to the former three year term to provide more stability for its research scientists.[27]

What is the lesson to be learned from these situations? Universities have dealt for decades with the federal government and its agencies as major research sponsors. When the government program directions changed this was done gradually—no dramatic changes generally took place to interrupt research. Government agencies normally provide stable, long term research support. Thus, the university does not concern itself too much with the possibility of sudden unexpected changes when dealing with the federal government.

In their initial research relations with industry, universities did not concern themselves with the cost effectiveness of corporate investment in university research. The demise of these two sponsors shows that decision making in corporations can unexpectedly turn off the spigot of industrial money. This can be done rather quickly. Universities will have to get used to the risk of quick shifts in research emphasis as a direct result of a change in the industry's marketing interests. Speculative research investment is constantly shifting so support is vulnerable to change.

Tom L. Tolbert, Ph.D., Director of External Research and Development at Monsanto Company, presented the following observations regarding his corporation's research participation with Washington University:

The agreement with Washington University Medical School is easily the most effective collaboration in which I have had the pleasure of being involved. It, in effect, has added a new dimension to our previous concepts of what university/industry collaborations should comprise. The whole climate of the relationship is upbeat and constructive. And I believe both groups find it productive. Synergism is an overused term, but it does seem to apply here when projects are undertaken jointly. The Needleman study on atrial peptides is an example. And certainly we at Monsanto enjoy the intellectual contact. As a result, working together is the option often preferred by our scientists, even if it may mean more initial cost for the corporation; our management tends to agree because of what we see coming from such efforts.

The general feeling at Monsanto now in working with Washington University scientists is that these relationships are not really different than working with a research associate, consultant, or related company. Our positive experience and the continuity of past relationships with the University have made us comfortable with the concept and with the idea that we can work out any issues which may arise. If personalities and interests are compatible, the rest is fairly straight forward. While concerns for the protection of academic freedom and the ownership of patent rights are still vitally important to each partner, these are now regarded as merely areas in which details have to be worked out. Monsanto understands fully that its needs will not be met if those of the University are not met. Likewise, Washington University recognizes Monsanto's needs and attempts to respect them. While it may sound like "flag waving," the relationship really is one of friends working together with mutual respect for each other's needs.

In the area of trade secrets and proprietary information, Monsanto is cognizant of the nature of the academic environment and the need for intellectual exchange. Monsanto therefore is willing to forego some of the controls and guarantees common in industry. There is a risk that proprietary data might be compromised, but this is a risk we can live with. We only ask that the university and its researchers, as professionals, exercise their "best effort" to keep data confidential. When an individual researcher is given proprietary data, the individual is informed of that and asked to keep it separate and confidential. Confidentiality depends on that individual's ethics in the last analysis, as always, and we must be willing to take the chance.

The contract itself serves as a guideline or statement of goals of the research relationship. Speaking from the investment end of the corporation, as opposed to the philanthropic, I can assure you that Monsanto is very pleased with the outcome of the research generated by the agreement. Monsanto is expanding its interests in this relationship and sees a definite return on its investment for the future.[28]

FUTURE ISSUES

The rules of contracting have made for successful research ventures between industry and institutions of higher education. No serious legal or cooperative issues appear to have manifested themselves. Mutual respect between the parties and good communication appear to have avoided any legal difficulties. Contract law, state and federal statutes, and self-regulatory guidelines appear to be successful in that no evidence appears that any draconian measures or law suits have been employed by the parties. The major reason for the potential non-renewal of the Mallinckrodt and Monsanto Hybridoma agreements has resulted from the shift in corporate research interests, not from dissatisfaction with the research results or perceived failure of either party to perform according to its contractual commitments.

From the viewpoint of the federal government, future concerns will center around foreign technological competition relative to the status of research and development in the United States. The United States is currently regarded as the world leader in both basic science and the commercial development of new biotechnology. Assurance of American corporate preeminence in commercialization of new technology is a primary concern of Congress. Future questions by Congress will look to how foreign interests are utilizing American universities for research, to what extent tax payers' dollars may be assisting in funding this research, and what steps need to be taken to maintain this country's competitive edge.[29]

This concern can be recalled by the following dialog before the House Subcommittees on Investigations and Oversights and Science, Research, and Technology when Congressman Gore was questioning the Hoechst agreement with Massachusetts General Hospital:

Mr. Gore: Will Federal money be used in the new department other than through the employees recruited from the hospital generally who have been trained by Federal money? Is that the only link with Federal money?

Dr. Lamont-Havers: That would be the only link that we could see at the present moment. It is visualized in the future that it is possible that in interactions and collaborative work with others, that there might be Federal funds. . . .

Mr. Gore: You don't think there is any other reason for concern at all? I mean, I am charged as a member of this committee with looking at the authorization of money for research in this country. And, thankfully, we allocated a lot to research, we ought to allocate more.

But my constituents have concerns about it, as most Americans do, and these are not easy questions, I understand that. But here you have an enormous amount of money from the taxpayers to this research institution and you have a continuous flow back and forth between the main body of the research institution and the genetic engineering part of it. The genetic engineering part of it is funded by a foreign chemical company and they have exclusive licensing rights for the ideas leading to products that come from the genetic engineering part of it.

Dr. Lamont-Havers: Which they funded.

Mr. Gore: Which they funded and which works in consert with the publicly funded portion of it. Sixty-five percent of it is coming from the public.

Dr. Lamont-Havers: Right.[30]

George Pake, former Provost of Washington University and currently Group Vice President for Corporate Research at Xerox Corporation, argued for a national rededication to United States technological leadership in his 1983 Ferguson Lecture. He emphasizes the need to improve and accelerate investment in research and development to meet foreign competition.

The United States annual basic research investment, from all sources, expressed in constant 1972 dollars has grown little since 1968.For a growing nation that is not encouraging. . . .

Six percent of all American degrees awarded are in engineering as compared to 37 percent of all West German degrees and 21 percent of all Japanese degrees.

With half our population, Japan graduates 5,000 more electrical engineers each year than we do. . . .

Some change is taking place. Recent bold new joint-research efforts between corporations and research universities have emerged. . . . But, along with a number of my colleagues from industry, I must stress that these special relationships are no panacea for national needs.

In terms of direct industrial needs, the essential research is *applied* research. Industry may fund basic research to provide a "window on the world" of new basic knowledge with commercial potential or to meet an obligation it may feel to replenish the store of basic knowledge. But industry's support of basic research will never be major because its commercial applicability is never predictable and often far in the future. . . .

Leadership in E, R & D must become our primary national goal. . . .

Renewed investment toward achieving technological leadership should be the first priority on the national agenda.[31]

The Federal government continues to support technological leadership in research by promulgating favorable legislation. At the conclusion of the 98th Congress, Second Session, at the end of 1984, two statutes were adopted which could enhance university–industry research agreements by reducing the potential of antitrust liability of participating corporations. The National Cooperative Research Act of 1984 (Pub. L. 98–462) clarifies the legal status of joint research and development ventures and encourages their formation.[32] The Local Government Antitrust Act of 1984 (Pub. L. 98–544) responds to the Supreme Court's decision in *Lafayette v. Louisiana Power & Light Co.*, 435 U.S. 389 (1978), and *Community Communications v. Boulder*, 455 U.S. 40 (1982), cases in which antitrust immunity was not automatically assured for local government entities acting under general home rule authority. It is arguable that a government unit as defined by this statute for protection could include state or regional R&D organizations as well as state boards or regents of higher education or similar political subdivisions established to conduct general or special research.[33] Lastly, the research and development tax credit established by Congress in 1981 is due to expire in 1985.[34] In all probability, Congress will renew and extend this credit which has potential benefits for university–industry relationships.

Government policies, however, do raise a variety of conflicting and diverse questions of which industry and university researchers should be aware. Donald Kennedy of Stanford University notes that the executive branch, Congress, the universities, and scientists are all in agreement that effective support of science is highly desirable. But declining financial support has had negative results which threaten to change the United States' way of doing science. The increasing competition for financial resources has caused three results:

The first is an increased attentiveness, on the part of the universities and investigators alike, to private sources of funds as a substitute for federal ones. The second is a disturbing incipient collapse of the agreement between the performers and the sponsors of research, that peer review is the fairest and most effective way of awarding scarce resources. The third . . . is a falling-out between the proprietors and the doers of scientific work over how the research venture is to be supported.[35]

It appears clear that a successful role for this country's technological leadership will be contingent upon the cooperation of all partners in the research and development system: industry–government, university–government and industry–university. Foreign competition will be a future concern for American industry, universities, and government. Continuation and expansion of the industry–university partnership has the promising potential of offsetting the impact of foreign technological competition on this country's economy and role in scientific leadership.

CONCLUSION

Alternative means for providing industry support to universities will begin to manifest themselves as various methods are explored for extending present research relations. Continued lack of accelerated federal funding for research and development in higher education coupled with industry's needs and the successes of existing contractual research arrangements will result in novel proposals. Present trends suggest that the "research and development limited partnership" (RDLP) offers a promising means of enhancing the ability of a university to exploit the commercial potential of its technological developments. Attorneys Joseph W. Bartlett and James V. Siena address this potential as follows:

Furthermore, it [RDLP] creates the possibility of allowing the university to share substantially in the profits from any product successfully developed. . . . [T]he tax laws make the RDLP an attractive investment vehicle for venture-capitalists and other investors. It is a device which can be used by a going business to finance research and development without demand on company finances. It can also be used by those interested in pursuing the commercial potential of an idea, leaving to a later time the decision on who will make and sell the product if the idea proves out.[36]

The RDLP developed directly from our federal tax laws, specifically Internal Revenue Code § 174(a). Investors are attracted through income tax benefits for funds which they provide to support research and development projects. Again, making reference to lawyers Bartlett and Siena we find the RDLP defined thus:

An RDLP consists of a general partner or partners and a limited partner or partners who join together to manage and finance the development of new technologies.

In many RDLPs, the general partner contributes the technology to be explored and the management expertise necessary to the success of the enterprise. The limited partners invest cash. For a variety of reasons, the university should not be the general partner. It could, however, choose to participate in the partnership either as a limited partner or through a licensing contract. In either case, its "investment" would take the form of a contribution of the technology to be explored.[37]

The benefit to the investors rests with their investment being largely deductible in the year it is made. Limited partners offset their income and diminish their marginal risk. The potential for financial returns through marketing successful research ventures is the ultimate reward for the investor.[38]

The benefit for the corporate researcher rests in the corporation's ability to finance research and development without dipping into the corporate treasury, issuing new stock, or increasing the company indebtedness.[39] If the university should elect an RDLP to develop and market its new technology, Bartlett and Siena suggest a checklist of issues which should be evaluated in creating an RDLP:

—should a university pool all its technologies and contribute them to a single RDLP;

—who should serve as general partner and will they bring the necessary expertise and resources;

—should the university receive an interest as a limited partner in return for its contribution of its technologies or should it receive a fee or royalty as a licensor,

—what standards of performance will insure expeditious development of marketable products;

—should reporting requirements for the general partner be stated and should an arbitration procedure be utilized to resolve disputes;

—what limitations will be placed on the use of the university's name;

—would an impartial advisory board of scientists, businessmen, and educators be useful in monitoring research developments.[40]

In proposing that Washington University look to the RDLP, Edward L. MacCordy, Associate Vice Chancellor for Research, and his assistant Harry S. Leahey identify four basic elements essential to the limited partnership:

First, there would continue to be an advisory committee which would operate under criteria set forth in the research agreement and which would have representation to protect the interests of the partnership. Approved projects would be carried out in the same manner as under present contracts.

Second, by carefully determining the optimum amount of funds to be raised, the research areas to be included, and the number of years over which the support should be available, this long term funding would allow the University to maintain its current excellence, attract new staff, and expand in areas which may now be limited or excluded.

Third, both Mallinckrodt and Monsanto could still participate as first preference licensees if desired. They would be relieved of having to finance the research, yet could have an exclusive option to license the tangible research results from the partnership in their fields of interest. Depending on the basic plan of operations, the partnership might also provide funding for clinical evaluation. It is even conceivable that funding for commercial development might be provided to licensees in order to bring products to the marketplace quickly.

Fourth, company licensees would pay royalties to the partnership which would then flow to the Limited Partners and to the university.[41]

The following list summarizes the advantages and disadvantages to the RDLP.

Advantages:

1. Research projects are conceived and conducted in the normal university manner independent of outside intervention.
2. A time period for assured funding of the program over five or possibly more years is feasible.
3. The scientific areas eligible for funding are determined by what may produce commercially viable results *not* necessarily limited to hybridomas or solely to the fields of interest of one or two companies. Nor is the distribution of funding among the areas based on the desires of these companies even though they may have a right of first refusal on research results from all or certain areas.
4. The amount of funding to be provided to the program can be matched to our potential to productively use it, including possible personnel acquisitions, rather than being limited to a company sponsor's ability and willingness to invest.
5. There could probably be greater freedom to fund lab improvements, equipment, etc., just like scientific personnel acquisitions, based on the concept of strengthening a productive research capability over a longer period than three years.

Disadvantages:

1. By initiating a concept of this sort in concert with the prospective General Partner, one assumes an ethical obligation to the scientifically naive Limited Partners. In dealing with corporate sponsors one assumes little if any responsibility for judging whether the risk/reward ratio of their present R&D investment in us represented a prudent investment. Unless one has reasonable confidence that RDLP which is initiated offers investors not only tax benefits but a good, albeit high risk, *potential* return on their investment, one should not instigate the RDLP. While the willingness of knowledgeable and relatively conservative established companies to invest heavily in such R&D should give us some degree of confidence, a general speculation must be made concerning the potential market value of the set of potential products which the university's R&D might reasonably create.
2. Advantages must be sufficient to offset a potential lower future royalty return compared to present contracts. Since royalty income from licensees flows back to the partnership and must be shared with the Limited Partners, the university's

effective royalty rate on sales could well be only 50% of the present contract rates. Maybe this would be offset by more effective marketing, ie., greater sales volume of a wider product mix.

The federal government's support of RDLPs has been stated before. It has made available its own guidelines and sample forms.[42]

Dangers and risks with RDLPs have been identified. These include sudden cancellation of an entire project, bankruptcy of the major corporate sponsor, and questions raised by the Internal Revenue Service as to the legality of the partnership based on perceived flagrant abuses by partnership sponsors' promising excessive write-offs or unusually high returns. Universities will be wise to know thoroughly the reputation and expertise of the general partner before entering into an RDLP.[43]

RDLPs also present other business risks. A concern first and foremost in the institution's mind is that the research may prove unsuccessful. No guarantee exists that a proposal or model can be commercially developed. A secondary concern recognizes a reality of high technology in that the threat of technical obsolescence is omnipresent. Competing research interests may with great rapidity successfully develop a better device or technique, thus destroying the RDLPs market.[44] Lastly, an RDLP may benefit from an innovative product only to find that the corporation does not have the financial or managerial expertise to market and promote the invention. This caution has primary application to research affiliations with small industries.[45]

Edward L. MacCordy makes the following observation:

A major question involving R & D LP's involves who will indemnify the university against marketplace risks. On the horizon universities may be dealing with a wider scale with smaller high tech corporations that are capital poor. These new corporations are not conservative or slow in marketing their products. Young high tech corporations have motivation and incentive to get products to the market. They have less corporate bureaucracy and place less emphasis on risk avoidance. Large corporations are content with a longer term payoff. Limited partnerships industries will want payoffs within a short period of time and tend to look to, or even create, these new, aggressive, high tech companies as marketing outlets for products of research. A question of indemnifying the university for any public liability which may develop in the future has still to be addressed. These smaller corporations may not survive long after their initial marketing successes and may not be in existence at the time a future product liability law suit takes place which is based on these initial products.

Future trends for innovative funding of university research is quite speculative. One approach rumored to be in development is funding by large investment and financial houses using a large "fund bank" to finance selected university research which has commercial potential. Another rumor indicates that a major financial house is setting up a "blind" RDLP for two specific universities to draw on based on research proposals submitted to a decision making group. This concept anticipates research monies being allocated by the financial houses, with the advice of scientific consultants, for research projects which are to be proposed after the fund is established. One can anticipate that these new research relationships with private industry and private investors are just beginning.[46]

It appears that the RDLP will impact university research agreements as the next step in identifying alternative funding mechanisms which will encourage and expand industry–university research relationships. Certainly the federal government's support through various tax advantages to investors will not go unnoticed by academia and industry. Investment and tax risks exist but can be minimized by prudent and careful evaluation of proposals and a thorough investigation into the reputations of general partners.

Another trend predicted as the future alternative for institutions of higher education involves universities creating their own technology-transfer organizations in cooperation with private venture capitalists. Several relationships are being formulated through Walter Channing, general partner of CW Ventures, Inc., and attorney Joseph Bartlett of the New York law firm of Gaston, Snow, Beekman & Bogue. The appeal of university research as an investment opportunity for private investors rests in the perception of the universities' undeveloped technology as promising a higher return than investments in the overcrowded high-technology laboratory of private companies. Additionally, financial institutions with large investment capital will most likely find university research an appealing option. Current trends in this area have been identified by G. Thomas Gibson:

To date, only a few research institutions have actually set up tech-transfer units, let alone turned ideas into companies. But universities moving determinedly in that direction include:

- Baylor University, which nurtures hopes for commercializing its expertise in virology and cardiovascular techniques.
- The University of West Virginia, which would like to market its plastics and materials science research.
- The University of Delaware, which sees potential in its marine engineering capability.
- Case-Western Reserve University, which has expertise in programmable machine tools and other engineering and medical technologies.
- The University of Utah, which wants to sell its biomedical technology.
- The University of Pittsburgh, which has hopes for research in new materials, robotics, and even a heart pump.[47]

Questions still remain unanswered regarding technology transfer between corporate bodies and non-profit higher education institutions. Attorney Joseph Bartlett indicates two areas for potential problems which must be addressed in the future:

> The first is exclusivity—the right of investors to commercialize research they fund. Not surprisingly, investors want to control everything a department produces. Institutions want investors to receive only specific products. . . .
>
> The second problem . . . is attracting investors with staying power. Developing successful projects, Bartlett warns, "will be a long, drawn-out process."[48]

Other universities' experiences can highlight some of the dangers in developing research relations involving outside investment promoters. The experiences of the microbiology department of McGill University's Medical School have recently surfaced. Researcher Irving DeVoe and his associate Bruce Holbein developed a metal-extraction process that apparently could separate gold from mine wastes and could remove toxic metals from polluted water. They were offered investment support to develop commercially their process by Irving Kott who, unknown to them, had a history of questionable financial dealings and a stock fraud conviction. Alan Freeman reporting in the *Wall Street Journal* notes:

> Thus, a leading university unwittingly lent its prestige to an international stock promotion by a convicted swindler. . . .
>
> McGill administrators were sympathetic when the two professors suggested a novel twist to the university's patent policy. Instead of giving the school 20% of the profits, they proposed giving McGill stock in the company they planned to set up, which then was to have been DeVoe-Holbein, Inc., a U.S. company.
>
> In an internal memorandum, McGill's principal, or president, David Johnston, voiced concern. He wrote that "it will probably become public knowledge that McGill is involved in an unusually speculative venture in which some people may lose money." But he felt that accepting the shares was the best way to protect the university's interest in any future patent, and he recommended that McGill take them.
>
> In the end, McGill received 8 percent of the stock in DeVoe-Holbein, Inc., in which the professors held a 31 percent stake. McGill also agreed to rent part of the microbiology department's lab to the company.[49]

As a result of questionable stock offerings by Kott and potential legal problems, McGill University placed its stock in a blind trust and appointed a Montreal lawyer to investigate conflicts of interest and misuse of government research funds charges against DeVoe and Holbein. As of this writing tensions remain and the variety of legal issues are unresolved. This incident serves as an illustration of the dangers inherent in a university's entering into a research-and-development project whose funding base is speculative.

Additionally, university researchers may find profitable and marketable products resulting from joint relationships with industry slow in reaching the commercial market place. Particularly in the monoclonal antibody research area, these medical weapons have only trickled into the market. Reasons for this slowness have been identified by *Wall Street Journal* reporter Hal Lancaster. He notes that the complexities of the human body have caused researchers problems. Tumor cells to be treated by monoclonal antibodies differ from patient to patient. Several antibodies may be required for one patient. Economic problems also confront commercial developers. Production processes are primitive and costly. Questions also exist as to whether or not patent protection can be applied to protect the hordes of monoclonal antibodies whose characteristics vary slightly. Lancaster notes:

> Most problems can probably be solved but that will take time. International Development Inc., a market research firm, predicted in 1982 that monoclonal sales would reach nearly $2.5 billion in five years. Now, says the company's president, Kenneth G. Bosomworth, the report seems "just too optimistic."[50]

Another reason identified as a significant delaying factor in affecting the production and marketing of monoclonal antibody products involves a perception of excessive government regulation. Miriam Rozen states the problem as follows:

> One of the major roadblocks faced by the biotech firms is getting their technologically exotic products approved by the Food and Drug Administration—a process that typically takes five to seven years. Thus, in planning for the interim, they are choosing products that can travel through the regulatory maze more rapidly.[51]

As a result, the young corporations are developing less dramatic products than cancer or diabetes cures. These products are primarily diagnostic devices or animal pharmaceuticals utilizing the science of monoclonal antibodies. These interim products permit the company to establish a market presence while satisfying its investors that they are actually doing something productive and profitable.

The burgeoning research relationship between corporate research interests and universities appears to have established itself as an acceptable replacement for or supplement to government funding of research. Selina Hunt reporting for the *Times (London) Higher Education Supplement* quotes William Carey of the American Association for the Advancement of Science:

> According to William Carey, executive officer of the association, American research is no longer the simple trilateral alliance between academia, government and industry that it was in the immediate postwar decades. It has now evolved into a pyramid of partnerships ranging from traditional contracts between sponsors and individual academics to complex deals negotiated strenuously by whole institutions.[52]

As the examination of the contracted relationships illustrates, the experiment appears successful from both the viewpoints of the industry and university. Additional research relations are being negotiated and new alternatives are being explored. Concerns for academia still exist regarding the impact of private funding on the research autonomy of scientists and the conflict of interest that may occur if scientists regard their research results as trade secrets. Governmental interests encourage industrial support of university research through tax incentives and new patent laws. Congress, however, appears concerned over the potential problem of federal tax monies being used to support research which might be commingled with research funded by foreign corporations. The reality of existing relationships at this time appears to minimize these concerns. Actual experience under current contracts shows a uniformity of opinion regarding the positive benefits which have resulted. Even in those contracted research relationships where renewal is in question, the renewal issue reflects shifting corporate research interests and not the lack of success of the agreement.

A. Bartlett Giamatti, President of Yale University, summarizes the current status of these relations:

> The opportunities for cooperative research between universities and industries are very exciting and can redound to the benefit of society. . . . We should negotiate appropriate arrangements, openly arrived at, which can further our mission. The constant challenge for the university is to know in clear and principled terms how to cherish learning, and its pursuit; and how to assist in bringing the results of free inquiry to the rest of society for the good of the public.[53]

A special task force established by the Twentieth Century Fund and chaired by Robert L. Sproull of the University of Rochester has just released its report examining industry–university research relations. This contempory report notes the benefits of University agreements with business:

> Multimillion-dollar research agreements between universities and corporations are not destructive of university values and can be beneficial to science, as long as the academic institutions take steps to prevent conflicts of interest. . . .
>
> The panel concluded that although corporations may interfere inadvertently with academic freedom, they "have no desire to tamper with or curtail the freedom that is critical to intellectual inquiry." The report noted that because corporations are certain to protect their own interests, it is up to universities to take the initiative in devising safeguards against corporate intrusions.[54]

The future can be based on the past as prologue. Today's research and development agreements will increase and expand. New options will be developed such as research and development limited partnerships. Self-monitoring by the Association of American Universities and individual university faculties will aid in the continued success of these ventures. Our national interests, society's betterment, and the university's role of being in the forefront of research assure the continued development of industry–university research relationships.

NOTES

1. U.S. Congress, House, Committee on Science and Technology, *University/Industry Cooperation in Biotechnology, Hearings before the Subcommittee on Investigations and Oversights and the Subcommittee on Science, Research, and Technology*, 97th Cong., 2d sess., 1982, pp. 4–5.

2. Barbara Culliton, "Reactors," in *Partners in the Research Enterprise: University-Corporate Relations in Science and Technology*, ed. Thomas W. Langiff et al. (Philadephia: University of Pennsylvania Press), p. 163.

3. Ibid., p. 162.

4. Ibid., p. 164.

5. Richard R. Nelson and Richard N. Langlois, "Industrial Innovation Policy: Lessons from American History," *Science* 219 (February 1983):814.

6. Sumner Myers and Donald G. Marquis, *Successful Industrial Innovations: A Study of Factors Underlying Innovation in Selected Firms* (Washington, D.C.: National Science Foundation, 1969), p. 1.

7. Thomas W. Langfitt, "Epilogue," in *Partners in the Research Enterprise*, pp. 168–171.

8. See Barbara J. Culliton, "Stanford Doctors Try Consulting, Inc.," *Science* 217 (17 September 1982):1122–1123.

9. See Suzanne Perry, "Professors Can Land Corporate Sponsors for Research—If They Follow the Rules," *Chronicle of Higher Education*, 20 April 1983, p. 21.

10. "Trading Jobs in Academia and Industry Helps High-Technology Firms," *Wall Street Journal*, 26 June 1984, p. 1.

11. See Marjorie Sun, "Biotech Firm Biogen is Going Public," *Science* 219 (25 February 1983):938–939.

12. See U.S. Department of Commerce, General Counsel for Economic Affairs, "Information and Steps Necessary to Form Research and Development Limited Partnerships," November 30, 1982 (mimeographed).

13. Larry R. Thompson, personal letter, July 8, 1983. See also, David Walker, "British Universities Step Up Research Parks to Foster Ties with Technology Companies," *Chronicle of Higher Education*, 16 March 1983, p. 19. See also Ira Fink, "The University as a Land Developer," *Planning for Higher Education* 12 (Fall 1983):9–32.

14. Louis G. Tornatzky et al., *University–Industry Cooperative Research Centers: A Practice Manual* (Washington, D.C.: National Science Foundation, May 1982).

15. See Robert Reinhold, "Stanford and Industry Forge a Research Link," *New York Times,* 10 February 1984, sec. A, p. 22.

16. Adrian Arima, personal letter, July 15, 1983.

17. Thomas H. Moss, "New Partnerships a Bargain for Industry, a Boon for Colleges and Universities," *Chronicle of Higher Education*, 6 April 1983, p. 72.

18. Thomas W. Langfitt, "Epilogue," in *Partners in the Research Enterprise*, pp.

172–173. See also, National Commission on Research, *Industry and the Universities: Developing Cooperative Research Relationships in the National Interest* (Pasadena, Calif.: National Commission on Research, 1980), pp. 25–32.

19. University of California Systemwide Administration, Office of the President, *Interim Guidelines on University–Industry Relations: Issues and Recommended Responses* (3 November 1982). See also, David E. Sanger, "U. of California Puts Limits on Private Research Pacts," *New York Times*, 21 August 1983, p. 12; Colin Norman, "Potential Conflicts of Interest Detailed at UC," *Science* 219 (21 January 1983):268–269.

20. American Civil Liberties Union, "Policy Statement of American Civil Liberties Union on University and Contract Research with Emphasis on Growing Ties between Corporations and Academic Institutions," 28 October 1983 (mimeographed).

21. Stanford University, *Policy on Tangible Research Property* (16 March 1982); Hubert Heffner, *Conflicts of Interest Situations* (31 August 1982); Gerald J. Lieberman and Robert L. Street, *Policy and Procedures Summary [for Sponsorship of Research]* (31 August 1982); Stanford University, *Policy on Consulting by Members of the Academic Council: Principles and General Standards* (2 December 1982).

22. Robert M. Rosenzweig to William Danforth, 10 January 1984.

23. Thomas H. Moss, "New Partnerships a Bargain for Industry, a Boon for Colleges and Universities," p. 72.

24. Jack Duvall, "Letter to the Editor," *Chronicle of Higher Education*, 18 May 1983, p. 32.

25. Interview with Harry W. Orf, Massachusetts General Hospital, Boston, Mass., 10 April 1984. See also, Katherine Bouton, "Academic Research and Big Business: A Delicate Balance," *New York Times Magazine*, 11 September 1983, p. 62.

26. Interview with David Kipnis, Washington University School of Medicine, 16 April 1984.

27. Interview with Edward L. MacCordy, Washington University, St. Louis, 29 February 1984.

28. Interview with Tom L. Tolbert, Monsanto Company, St. Louis, Missouri, 25 June 1984.

29. See, generally, U.S. Congress, Office of Technology Assessment, *Commercial Biotechnology: An International Analysis* (Washington, D.C.: U.S. Government Printing Office, 1984). See also, U.S. Congress, House, Committee on Banking, Finance and Urban Affairs, Subcommittee on Economic Stabilization, *Forging an Industrial Competitiveness Strategy*, 98th Cong., 1st sess., 1983, Committee Print 98–8, pp. 16–20; U.S. Congress, House, Committee on Foreign Affairs, *Overview of U.S. International Competitiveness, Hearings before a Subcommittee of the House Committee on Foreign Affairs on Aerospace Industry, Machine Tools Industry, and Electronics Industry*, 97th Cong., 2d sess., 1983, pp. 139–141.

30. U.S. Congress, House, Committee on Science and Technology, *Commercialization of Academic Biomedical Research, Hearings before the Subcommittee on Investigations and Oversights and the Subcommittee on Science, Research, and Technology*, 97th Cong., 1st sess., 1981, pp. 94–95.

31. George Pake, "U.S. Technology in Jeopardy," *Washington University Magazine* 54 (Spring 1984):18–23.

32. 98 Stat. 1815 (1984). See William F. Baxter, "Antitrust Law and Technological Innovation," *Issues in Science and Technology* (Winter 1985):80–91.
33. 98 Stat. 2750 (1984).
34. Laura Saunders, "The R&D Dilemma," *Forbes* 135 (25 February 1985):130.
35. Donald Kennedy, "Government Policies and the Cost of Doing Research," *Science* 227 (1 February 1985):481.
36. Joseph W. Bartlett and James V. Siena, "Research and Development Limited Partnerships as a Device to Exploit University Owned Technology," *Journal of College and University Law* 10 (Spring 1983–84):437.
37. Ibid., p. 438. See also, Patrick J. Ellingsworth, "Investing in Research and Development Limited Partnerships," *National Law Journal* 5 (18 July 1983):38.
38. See Nicholas G. Moore, Jay F. Ayers, and Frank R. Pope, "How Limited Partnerships Tax-Shelter the R&D of New Products or Technology," *Journal of Taxation* 49 (September 1978):138.
39. See Michael Waldholz, "R&D Partnerships Proliferate, Offer Investors a Tax Shelter Plus a Shot at High-Tech Profits," *Wall Street Journal*, 26 September 1983, sec. 2, p. 60.
40. Joseph W. Bartlett and James V. Siena, "Research and Development Limited Partnerships as a Device to Exploit University Owned Technology," pp. 439–442.
41. Edward L. MacCordy and Harry S. Leahey to Dr. Joseph Davie, 18 August 1983.
42. See U.S. Department of Commerce, *Information and Steps Necessary to Form Research and Development Limited Partnerships* (November 1982).
43. See Thomas J. Murray, "The Pitfalls of R&D Partnerships," *Dun's Business Month* 123 (April 1984):127–128.
44. See Doren P. Levin, "Westinghouse Move into Robotics Shows Pitfalls of High-Tech Field," *Wall Street Journal*, 14 May 1984, sec. 2, p. 27.
45. See Patrick J. Ellingsworth, "The R&D Limited Partnership: When the Research Succeeds," *National Law Journal* 5 (25 July 1983):21.
46. Interview with Edward L. MacCordy, Washington University, St. Louis, 29 February 1984.
47. G. Thomas Gibson, "Research for Sale," *Venture: The Magazine for Entrepreneurs*, March 1984, p. 80.
48. Ibid., p. 86.
49. Alan Freeman, "How a Top University Unwittingly Became Stock-Promotion Bait," *Wall Street Journal*, 27 April 1984, p. 1, col. 6.
50. Hal Lancaster, "Efforts to Develop Monoclonal Antibodies Are Moving More Slowly Than Expected," *Wall Street Journal*, 1 May 1984, sec. 2, p. 37, col. 4.
51. Miriam Rozen, "Biotech Firms Offer 'Interim' Products," *Dun's Business Month* 123 (May 1984):103.
52. Selina Hunt, "Private v. Public Enterprise," *Times Higher Education Supplement*, 23 March 1984, p. 13.
53. A. Bartlett Giamatti, "Free Market and Free Inquiry: The University, Industry, and Cooperative Research," in *Partners in the Research Enterprise*, ed. Thomas W. Langfitt et al., p. 9.
54. Kim McDonald, "Safeguards Urged in University Agreements with Business," *Chronicle of Higher Education* 5 July 1984, p. 4.

Appendix A
Massachusetts General Hospital-Hoechst A. G. Contract of 1980

AGREEMENT, effective as of this 14th day of May, 1981, by and between THE GENERAL HOSPITAL CORPORATION, a not-for-profit corporation formed under the laws of the Commonwealth of Massachusetts ("MGH") and Hoechst AG, a corporation organized and existing under the laws of the Federal Republic of Germany (the "Company").

W I T N E S S E T H:

WHEREAS, MGH is engaged in ongoing programs of basic research in accordance with the scientific and educational purposes for which it was founded, and

WHEREAS, the Company desires to provide funding to MGH for the purpose of doing basic research in the Field of Research as defined in this Agreement, and

WHEREAS, MGH is willing to accept such funding from the Company in accordance with the terms and conditions provided in this Agreement,

NOW, THEREFORE, the parties hereto hereby agree as follows:

ARTICLE I

DEFINITIONS

1.1. Field of Research

"Field of Research" shall mean research in the field of molecular biology generally with specific initial emphasis in the areas of eukaryotic cell gene regulation, somatic cell genetics, microbial genetics, virology, immunology, and plant molecular biology.

1.2. Sponsored Research

"Sponsored Research" shall mean research funded in whole or in part by the Company pursuant to this Agreement.

1.3. Department

"Department" shall mean the separate laboratory or series of laboratories at MGH which shall be designated as the "Department of Molecular Biology", located in space more fully described in Exhibits A and B, to be established, staffed and administered by MGH in accordance with the provisions of this Agreement with funding provided by the Company for the purpose of doing basic research in the Field of Research in accordance with the scientific and educational purposes of MGH with a view to opening new avenues in the life sciences and providing a knowledge base for the development of new products of benefit to mankind.

1.4. Director

"Director" shall mean the scientist selected to head the Department in accordance with the provisions of this Agreement.

1.5. Senior Investigators

"Senior Investigators" shall mean the scientists selected to head research in the areas set forth in Section 1.1.

1.6. Scientific Advisory Board

"Scientific Advisory Board" shall mean the scientific board established pursuant to Section 5.2 hereof having oversight and review functions with respect to the work at the Department.

1.7. Invention

"Invention" shall mean any new and useful process, machine, manufacture, or composition of matter (a) conceived prior to this Agreement but first reduced to practice during Sponsored Research, or (b) conceived or reduced to practice as a result in whole or in part of Sponsored Research.

1.8. Patent Application

"Patent Application" shall mean any United States or foreign patent application, including any division, continuation or continuation-in-part thereof, claiming any Invention in which MGH has rights by virtue of sole or joint inventorship by a staff member, student or employee of the MGH or by the terms of a written agreement.

1.9. Patent

"Patent" shall mean any United States or foreign patent, including any reissue patent, issuing from any Patent Application.

1.10. New Research Building

"New Research Building" shall mean the new research building referred to in Exhibit B.

ARTICLE II

RIGHTS AND OBLIGATIONS OF THE COMPANY

2.1. Funding

The Company shall provide MGH with funds for the following purposes:

(a) funds for the design, renovation and furnishing of the initial space to be occupied by the Department on Jackson 11 and 12 and equipment for the Department;

(b) funds for the design and construction of space occupied by the Department in the New Research Building;

(c) funds for additional equipment for the Department in the New Research Building; and

(d) funds for the annual operating costs of the Department required to carry on Sponsored Research during the term of this Agreement;

all in accordance with the payment schedule provided for in Section 2.2.

All equipment and furniture purchased with funds provided by the Company shall be the property of MGH, and shall be maintained and insured by MGH for the purpose of the Department, in the same manner and to the same extent as comparable MGH property is maintained and insured (but with the expense thereof being charged as an operating cost of the Department hereunder). In the event that MGH desires to transfer such equipment and furniture out of the Department during the term of this Agreement, it may do so with the consent of the Director and upon paying to the Company the fair market value thereof determined at that time.

2.2. Amount and Timing of Payments

A. The Company shall provide guaranteed funding and may provide discretionary funding for the Department under this Agreement in accordance with the following schedule:

(a) Guaranteed funding in an amount not to exceed $2,300,000 to cover the costs of design and renovation of space at MGH (Jackson 11 and 12) and $2,500,000 to cover the cost of furniture and equipment therefor, to be paid by the Company with reasonable promptness upon receipt of appropriate evidence or notice that such payment is due;

(b) Guaranteed funding in an amount not to exceed $10,800,000 to cover the cost of designing and constructing the 30,000 square feet of space to be occupied by the Department in the New Research Building. It is expected that such funding shall be provided to MGH on or before

September 1, 1981; it may, however, be deferred at the option of the Company but shall in any event be paid on or before the commencement of construction of the New Research Building.

(c) Guaranteed funding in an amount not to exceed $2,000,000 to cover the costs of furniture and equipment for 30,000 square feet of space in the New Research Building, the same to be paid by the Company with reasonable promptness upon receipt of appropriate evidence or notice that such payment is due;

(d) Guaranteed funding in an amount not to exceed the following annual sums to cover annual operating costs of the Department required to carry on Sponsored Research during each of the years noted:

Fiscal Year Ending Sept. 30	Guaranteed Funds for Operating Costs (in addition to guaranteed funds for design, furniture and equipment for the Department)
1981	$ 500,000.00
1982	$ 3,600,000.00
1983	$ 3,600,000.00
1984	$ 6,000,000.00
1985	$ 6,000,000.00
1986	$ 6,000,000.00
1987	$ 6,000,000.00
1988	$ 6,000,000.00
1989	$ 6,000,000.00
1990	$ 6,000,000.00

Funding to cover operating costs pursuant to this Section 2.2A(d) shall be paid by the Company to MGH on a quarterly basis. At least five days prior to the first business day of each fiscal quarter during the term of this

Agreement, the Company will provide MGH with the amount estimated to be needed for such quarter. Promptly after the end of such quarter MGH will provide the Company with a statement of actual operating costs of the Department for such quarter, which statement will reconcile the estimated and actual costs. To the extent that actual costs in such quarter exceed estimated costs, the Company shall add the deficit to the next quarterly payment to be made by it. To the extent that estimated payments exceed the actual costs in such quarter, the Company may credit the excess against the next quarterly payment to be made by it. Nothing in the foregoing quarterly payment and reconciliation procedure shall increase or reduce the Company's annual funding obligations under this contract.

(e) Discretionary funding in such additional amounts as the Company, in its sole judgment, may approve after consultation with the Director and MGH, as appropriate.

B. (a) The guaranteed funding amounts provided for in Section 2.2A (as adjusted pursuant to Section 2.2C) plus any discretionary funding amounts from time to time hereafter approved under Section 2.2A(e) constitute the limits of the Company's financial liability under this contract.

(b) Should the actual costs described in Sections 2.1(a), 2.1(b), 2.1(c) and 2.1(d) be less than the funding guaranteed to cover such costs provided for in Sections 2.2A(a), 2.2A(b), 2.2A(c) and 2.2A(d) (as adjusted pursuant to Section 2.2C) the funding guaranteed shall be automatically reduced to such actual cost level.

(c) Prior to making any of the payments called for by this Section 2.2, the Company shall have the right to review a detailed budget prepared by MGH and submitted to the Company pursuant to Section 4.2.

(d) Payments under Section 2.2A shall be made by the Company to MGH, who shall be responsible for actual disbursement of the funds.

C. The guaranteed amounts set forth in Sections 2.2A(b) and (d) are in 1981 dollars. Such amounts shall be adjusted to reflect changes in the value of the dollar between 1981 and the time of payment as provided below:

(a) If the $10,800,000 payment for the construction of the New Research Building set forth in Section 2.2A(b) is not made by September 1, 1981, it shall be increased by 1% per month for each month or portion thereof which such payment is delayed beyond September 1, 1981.

(b) The guaranteed amounts set forth in Section 2.2A(d) for each of the fiscal years ending 1983 through 1990 shall be adjusted at March 31 in the fiscal year of payment by multiplying such amount by a fraction, the numerator of which is the CPI-Medical Care Professional Services Component (published by the U.S. Bureau of Labor Statistics) (the "Index") for the calendar year ended December 31, 1982 through 1989, respectively, and the denominator of which is the Index for the year ended December 31, 1981.

For purposes of this section the Index referred to above shall be the most recent such figures published by the agency responsible therefore at the time of adjustment, even though such figures may be tentative and subject to subsequent readjustment.

2.3 Reimbursement or Repayment in Certain Events.

A. Upon the termination or expiration of this Agreement:

(a) MGH shall pay to the Company the fair market value, determined at that time, of such equipment and furniture purchased with funds provided by the Company as MGH desires to retain; and

(b) MGH shall offer to the Company at no cost such equipment and furniture as MGH desires not to retain and to the extent the Company does not want to accept such furniture and equipment, it shall be sold or otherwise disposed of by MGH and the net proceeds (after expenses of sale) thereof turned over to the Company.

(c) MGH shall repay to the Company an amount (not less than zero) equal to the construction cost paid to the MGH by the Company under Sections 2.2A(b) multiplied by a fraction, the numerator of which is the number of years between the termination date and June 30, 2001 and the denominator of which is 20, provided however that in no event (other than a material breach of this agreement by MGH) shall MGH be obligated to repay the Company more than half of such construction cost.

B. In the event MGH does not commence construction of the New Research Building on or before April 1, 1982, MGH shall repay to the Company the amount which the Company paid to MGH under Section 2.2A(b) (increased as provided in Section 2.2C(a)) together with interest on such amount from the date of payment to the MGH until the date of repayment to the Company at the rate which MGH earned on such amount by investments in short-term obligations of the United States Government or certificates of deposit of major banks.

ARTICLE III

TERM OF THIS AGREEMENT

3.1. Initial Term

This Agreement shall have an initial term ending September 30, 1990. Thereafter this Agreement shall be automatically extended for additional periods of five fiscal years, upon the same terms and conditions and at the same funding level as in effect at the end of the term (subject to continued adjustment for changes in the Index), unless either party shall notify the other, in writing, prior to the end of the second fiscal year prior to such automatic renewal, that it elects not to renew this Agreement.

At the end of fiscal 1986, the joint Committee referred to in Article VII shall review the performance of the Department and report its findings. If, on the basis of this review, MGH and the Company are of the opinion that the purposes for which the Department was established are not being carried out in a satisfactory manner then, after consultation and with the approval of the Company, MGH shall take such steps as it believes necessary and proper to correct such situation; but such review and any consequent reorganization shall not in any way relieve either party of its obligations to the other for the continued support and operation of the Department and performance of this Agreement for the balance of the initial term of this Agreement.

3.2. Exclusive Funding Rights. During the term of this Agreement, the Company shall have the right to fund all research at the Department in addition to research the funding for which is guaranteed pursuant to Section 2.2A(d). Any such research the Company declines to fund may be funded by other sources; provided, however, that MGH shall not, without the prior written consent of the Company, seek or accept funding for such research to be conducted at the Department from any other profit-making entity; and provided, further, that, in the event that the Company shall notify MGH of its election not to renew this Agreement, MGH may during the final two fiscal years of this Agreement seek and accept funding from any source, such funding to commence not earlier than the termination of this Agreement unless otherwise agreed by the Company.

MGH agrees to do nothing in the renovating or the initial equipping of Jackson 11 and 12 or in the construction or initial equipping of the new laboratory space in the New Research Building to allow any third party, including the United States Government, to acquire any rights or equity in any work solely accomplished in the Department by personnel of the Department. MGH represents that to the best of its knowledge there is on the date hereof no state of facts relating to the construction, renovation or initial equipping of Jackson 11 and 12, or to the construction or initial equipping of the new laboratory space, that would provide any such rights to any third party, including the United States Government.

ARTICLE IV

RIGHTS AND OBLIGATIONS OF MGH

4.1. Space and Equipment of the Department

MGH will forthwith establish, staff and administer the Department to carry on Sponsored Research in accordance with this Agreement as a separate department of MGH. The Department will be initially located in the MGH space described in Exhibit A attached hereto (herein referred to as "Jackson 11 and 12"). MGH shall provide to the

Department scientific and support staff of approximately 50 persons and shall use its best efforts to increase the space within a period of three years to approximately 30,000 square feet of interior space by supplying space in the New Research Building and scientific and support staff of approximately 100 persons. The MGH shall, as soon as is reasonably possible, apply for a Massachusetts Health and Educational Facilities Authority bond issue to cover the costs of design and construction of the New Research Building.

All contracts for the preparation of space, the purchase or lease of equipment or furnishings, the hiring of staff and operation of the Department shall be in the name of MGH, which shall be responsible for the administration and activities thereof. MGH shall exercise the same care and prudence in the establishment and administration of the Department as it exercises with respect to other departments or operations of MGH.

4.2. Department Operating Budgets

Three months prior to the beginning of each fiscal year during the term of this Agreement, MGH shall prepare and submit to the Company for its review an annual budget for the operation of the Department during such fiscal year. Such budget shall be prepared in such detail as shall be reasonably requested by the Company and may include as operating costs of the Department allocations for indirect costs in accordance with Exhibit C attached hereto. Such budget shall separately identify each capital expenditure of $50,000 or more, which capital expenditures the Company shall not be obligated to fund.

In the event that the Company shall decline to fund research included in the budget but in excess of its guaranteed funding obligation, or shall decline to fund a capital expenditure in excess of $50,000, it shall so notify MGH within thirty (30) days after receipt of the budget, in which event MGH may seek other funding for such research or capital expenditure in accordance with Section 3.2.

4.3. Personnel

MGH shall, after consultation with the Company, select candidates for the posts of Director and five (5) Senior Investigators. Such candidates shall be selected on the basis of recommendations of a search committee appointed by MGH in accordance with its normal policies and procedures for the selection of senior staff. It is agreed that Dr. Howard M. Goodman shall be the initial Director. The Director shall have a rank and function equivalent to a Chief of Service at MGH. The employment of the Director shall be for a term of ten (10) years. The Director and the Senior Investigators shall be regular members of the staff at MGH, nominated for membership in the faculty of the Harvard Medical School, and as appropriate, recommended for tenure at the Harvard Medical School. MGH shall enter into written Participation Agreements with the Director, the Senior Investigators, other Department personnel and others engaged in Sponsored Research in the form attached hereto as Exhibit D. All conditions of employment for the Director, the Senior Investigators and others engaged in Sponsored Research shall be arranged in accordance with the normal practices of MGH.

Costs and expenses incurred in search and recruitment activities for the Department shall be deemed part of the operating costs of the Department.

ARTICLE V

THE DEPARTMENT

5.1. Administration

All of the Sponsored Research will be carried on through the Department, under the direction of the Director. The Director will be responsible for the coordination of the work of the Department and the preparation and submission of an annual written report to the Company, to MGH and to the Scientific Advisory Board of the Department setting forth the significant developments at the Department during the past fiscal year and outlining its direction for the coming year.

The Director and each Senior Investigator, with the approval of the Director, shall select subjects of research and formulate research programs which conform to the broad goals of the Department.

5.2. Scientific Advisory Board

A Scientific Advisory Board shall be established consisting of six (6) qualified scientists, two (2) of whom may be affiliated with and shall be appointed by the Company, two (2) of whom may be affiliated with and shall be appointed by MGH, and two (2) of whom shall be unaffiliated with either the Company or MGH, but shall be jointly appointed by the Company and MGH.

The Scientific Advisory Board shall be responsible to review the annual report prepared by the Director and shall prepare its independent evaluation of the work of the Department. It may consult with the Director and with Senior Investigators of the Department, where appropriate, and will make recommendations concerning the work and operation of the Department either on its own initiative or in response to a request from either the Company or MGH. Recommendations made to the Department by the Scientific Advisory Board shall be deemed to be advisory only.

5.3. Faculty Duties

The Director and each Senior Investigator will devote their time primarily to research for the Department within the Field of Research and related activities. The Director and each Senior Investigator may also devote a reasonable amount of time to faculty duties other than research and to consulting for non-profit-making entities so long as such activities do not interfere materially with their research activities under this Agreement.

5.4. <u>MGH Rules and Regulations</u>

Except as otherwise explicitly provided herein, the Department and its staff shall be subject to all rules, regulations, policies and procedures of MGH. Research conducted at the Department or by Department personnel will conform to MGH research and experimentation regulations and to all federal, state, and local laws applicable thereto.

5.5. <u>Senior Investigator Reports</u>

The Senior Investigators shall prepare individual annual reports on the progress of their research. The Director shall incorporate such individual reports in the annual report described in Section 5.1. Each Senior Investigator's report shall include reprints of all scientific articles published by such Investigator during the year covered by the report.

5.6. <u>Investigator Status at MGH</u>

The Director and the professional staff of the Department shall have the same access as other members of the MGH professional staff to the facilities of MGH.

5.7. <u>Scientific Programs and Reports for the Company</u>

At least once a year, the Department will present a symposium of two or three days' duration for invited members of the academic community for the purpose of discussing the programs and research being conducted by the Department. The Company may send its employees and other individuals to such symposium, but shall provide the Department notice of the number of such people to attend. In addition, the Director will report directly to the Company's representatives up to three (3) times a year in Frankfurt, Germany or such other place as shall be designated by the Company. Senior Investigators of the Department shall be available to confer with representatives of the Company at least once a year in Frankfurt, Germany or at such other place as shall be designated by the Company.

5.8. Training

The Company shall have the right to send up to four (4) individuals to work and be trained at the Department at any one time. These individuals shall have qualifications acceptable to the Department.

5.9. Collaborative Work

(a) Each scientist at the Department shall be free to collaborate with others, subject to the terms of his Participation Agreement.

(b) Research collaborations funded in part by the Company and in part by others shall take into account the interest of the Company in obtaining exclusive, world-wide licenses. Such collaborations shall be arranged in a manner which will entitle the Company to the most favorable License (as defined in Section 6.3) obtainable, which shall be at least a nonexclusive license.

(c) MGH, without the prior written consent of the Company, shall not enter into any agreement for research collaboration to be conducted by employees of the Department with any profit-making entity.

5.10. Office of Technology Administration

MGH, acting through its Office of Technology Administration, shall maintain, administer and use its best efforts, to enforce such policies and procedures, including specifically the Participation Agreements, the Statement of Policy on Consulting Arrangements and the Patent Policies and Procedures set forth as Exhibits D, E and F to assure compliance with this Agreement and protect the interests of the Company hereunder.

ARTICLE VI

PROPERTY RIGHTS

6.1. Right to Publish

The right of individual scientists employed at the Department to publish research results in accordance with the educational and scientific purposes and policies of MGH shall not be infringed. MGH will submit to the Company early drafts of all manuscripts authored by members of the Department resulting from any Sponsored Research not less than 30 days prior to the submission of the manuscript for publication and the Company shall have the right to advise MGH as to the patentability of any inventions disclosed therein. At the end of such thirty-day period, the scientist shall have the right, at his sole discretion, to submit such manuscripts for publication.

6.2. Patent Rights

Each individual working at the Department or otherwise participating in Sponsored Research shall promptly disclose to the MGH Office of Technology Administration all Inventions arising out of such Sponsored Research in accordance with his Participation Agreement. The Office of Technology Administration shall promptly advise the Company of all such Inventions. Representatives of MGH and the Company shall then discuss whether Patent Applications shall be filed with respect to such Inventions and, if so, the scope of such Applications. In the event the Company and MGH agree that Patent Applications should be filed, the Application shall be filed in the name of MGH, but at the expense of the Company. In the event the Company is not interested in having Patent Applications filed with respect to a particular Invention, it shall advise MGH of such fact within ninety (90) days of being advised of the Invention by MGH and MGH shall be free to do so at its own expense or may dispose of the patent rights to such invention, or release them to the inventor, as it deems fit. In the event that MGH does not wish to file a Patent Application with respect to a particular Invention, or does not wish to file Patent Applications with respect to specific countries, it shall without delay notify the Company and the Company shall be free to file patent applications in its own name, and MGH shall render the Company all necessary assistance, including assignment of patent rights, in order to facilitate such filing. All costs of mutually agreed-upon prosecution, maintenance, working and defense of Patents licensed exclusively to the Company shall be borne by the Company. In any case where the Company is granted only a nonexclusive license, the Company will share the costs of patent prosecution, maintenance, working and defense with the other nonexclusive licensee or licensees. The Company shall be free to make commercial use of any nonpatentable discovery arising out of the Sponsored Research.

MGH, acting through its Office of Technology Administration shall maintain, administer and, to the extent practical, enforce such policies and procedures as are reasonably necessary to comply with this Agreement. MGH shall cooperate with the Company, at the expense of the Company, in the maintenance, working, prosecution and defense of any Patent Application or Patent licensed to the Company hereunder.

6.3. <u>Licenses</u>

A. As to each Patent Application which MGH shall file during the term of this Agreement, MGH shall, upon written request, submitted by the Company within the eighteen (18) months next following the date of such Application, grant to the Company and any affiliated entity designated by the Company, a license as defined in the following paragraph B, including whenever possible the right to grant sublicenses thereunder; and, in the absence of such request, MGH may grant a license to such Patent to any other person or persons on any terms.

B. The license granted pursuant to 6.03A ("License") shall be:

(a) with respect to Patents resulting from Sponsored Research funded exclusively by the Company, an exclusive world-wide license for the life of the Patent;

(b) with respect to any Patent resulting from collaborative research funded in part by the Company, an exclusive world-wide license for the life of the Patent whenever possible, and when not possible the most favorable license obtainable but in any event a nonexclusive world-wide license for the life of the Patent;

(c) with respect to any Patent claiming an Invention conceived during the term of this Agreement as a result of Sponsored Research but first reduced to practice within the 30 months next following the termination of this Agreement, the most favorable license obtainable but in any event a nonexclusive world-wide license for the life of the Patent.

C. With respect to any Invention conceived prior to affiliation of the inventor with the Department as to which any other person or entity may have rights, it is understood that any License to the Company with respect to such Invention may be subject to such rights.

D. In the event that the Company, either directly or through an affiliate or sublicensee, does not begin actual commercial development of any Invention (i.e., make use of the Invention with the bona fide intention to market products as soon as practicable) licensed or sublicensed to such entity under this Agreement within three years after the date of filing of the Patent Application with respect thereto, any license or sublicense granted to any such entity hereunder for such Invention shall become nonexclusive.

6.4. License Royalties

The Company shall pay MGH royalties for any License granted pursuant to Section 6.3 of the Agreement. The royalty rates shall be established giving due consideration to the Company's funding of Sponsored Research and, during any period in which the difference between

(x) the aggregate amount of royalties that would have been paid by the Company through such period on all licenses had the royalty rates been the fair commercial rates, and

(y) the aggregate amount of royalties actually paid by the Company through such period on all licenses

is less than the aggregate amount of all payments made by the Company under Sections 2.2A through such period, all calculated on a cumulative basis, the royalty rates paid by the Company shall not exceed, on any, license 50% of the fair commercial royalty rate for such license. In the event the parties cannot agree on the fair commercial royalty rate, the matter will be submitted to arbitration in accordance with the procedures of the American Arbitration Association.

With respect to any Patent as to which the Company and its designated affiliates are not the only licensees, the royalty rate to the Company and its designated licenses shall be no less favorable than the royalty rate to any other nongovernmental licensee under such Patent. Any royalties paid by the Company hereunder will be allocated among the inventor, the Department, the inventor's laboratory, and the general research funds of the MGH in accordance with the Patent Policies and Procedures set forth in Exhibit F.

The portions of any royalties allocated to the Department and to the inventor's laboratory shall be used to pay operating expenses of the Department, in which case the Company's obligation to provide guaranteed funding pursuant to Section 2.2A(d) shall be reduced by an equal amount.

ARTICLE VII

Joint Committee

The MGH and the Company shall form a Joint Committee composed of three members of the Board of Trustees of the MGH and three senior executives of Hoechst AG designated by the board of management of Hoechst AG. The function of the Joint Committee shall be to oversee the implementation of this Agreement and serve as a forum for communication between the MGH and Hoechst AG with respect to matters arising under the Agreement. The Joint Committee shall meet at least once each year at locations to be determined by the Committee.

ARTICLE VIII

ASSIGNMENT

Either party may assign its rights and obligations under this Agreement to any corporation controlling, controlled by or under common control with the party, provided however that:

(a) Any assignment by the Company shall not be effective unless accompanied by a guaranty by the

Company of the performance of its assignee hereunder; and

(b) Any assignment by MGH shall not be effective unless accompanied by a guaranty by MGH of the performance of its assignee hereunder and shall not operate to change the scientific character of the Department or to give rights in Sponsored Research to any profit making entity.

ARTICLE IX

NOTICES

9.1. Notice to the Company

All notices to the Company under this Agreement shall be deemed effective if made in writing and deposited in the United States Post Office, or with the postal service of the Federal Republic of Germany, postage prepaid, address as follows:

Dr. Hansgeorg Gareis
Hoechst Aktiengesellschaft
Postfach 80 03 20
6230 Frankfurt (Main) 80
West Germany

With copy to:

William H. Griesar, Esq.
Rogers Hoge & Hills
90 Park Avenue
New York, New York 10016

9.2. Notice to MGH

All notices to MGH under this Agreement shall be deemed effective if made in writing and deposited with the postal service of the Federal Republic of Germany or in the United States Post Office, postage prepaid, addressed as follows:

General Director
The General Hospital Corporation
Fruit Street (White 1)
Boston, Massachusetts 02114
U.S.A.

With copy to:

David M. Donaldson, Esq.
Ropes & Gray
225 Franklin Street
Boston, Massachusetts 02110
U.S.A.

ARTICLE X

BINDING EFFECT

This Agreement shall be binding upon and inure to the benefit of the parties hereto and their respective successors and assigns. However, the parties agree that the obligation of the Company to go forward with this Agreement is contingent upon Howard Goodman making an arrangement with Advanced Genetic Sciences, Inc. ("AGS") to protect Howard Goodman, the MGH and the Company from claims made by AGS. In the event such an arrangement is not made to the satisfaction of the Company or or before September 1, 1981, either party may terminate this agreement by giving notice to the other party, in which case any sums heretofore paid hereunder shall be returned.

ARTICLE XI

ENTIRE AGREEMENT

This Agreement constitutes the entire agreement between the parties with respect to the funding and exploitation of research in the Field of Research and supersedes all prior agreements, whether written or oral, and shall not be modified except by a writing signed by both parties.

ARTICLE XII

GOVERNING LAW

This Agreement will be governed by and construed in accordance with the laws of The Commonwealth of Massachusetts.

ARTICLE XIII

MISCELLANEOUS

The headings in this Agreement are for convenience of reference only and shall not alter or otherwise affect the meaning hereof.

IN WITNESS WHEREOF, the parties have caused this Agreement to be executed by their duly authorized representatives as of the day and year first above written.

THE GENERAL HOSPITAL CORPORATION

DATED: May 14, 1981 By____________________

HOECHST A.G.

DATED: May 14, 1981 By____________________

EXHIBIT A

Proposed Interim Research Facilities for the Department of Molecular Biology

The Department of Molecular Biology will occupy the renovated floors of Gray/Jackson 11 and 12. This represents 11,204 net usable square fee of space. There will be six large, open laboratories of approximately 800 - 1,000 square feet. Each lab will accommodate six to ten investigators. Two additional, smaller laboratories for bacteria reseach will be designated P-2 for cultures of viable organisms containing recombinant DNA molecules. One tissue culture lab for four researchers on plant materials will be designated P-3 for cultures of viable organisms containing recombinant DNA molecules. A conference room/library will have the flexibility to handle meetings of up to 25 people at a table or 50 people in rows of chairs for seminars. One kitchen/glasswashing room will take care of the clean and soiled glassware for both floors. Support facilities on each floor will include a darkroom, warm and cold rooms and an equipment room for high-speed centrifuges and freezers. Space will be available on each floor for electrophoresis and gel-running. There will be a total of fifty (50) people working on both floors. In an adjacent building there will be a greenhouse for the controlled propagation of plant materials.

EXHIBIT B

Proposed Permanent Research Facilities for the Department of Molecular Biology

The New Research Building will consist of a new building of not more than ten floors to be built by MGH on the MGH grounds and financed by a loan obtained by MGH from the Massachusetts Health and Education Facilities Administration. Upon completion of the New Research Building, the Department of Molecular Biology will occupy the upper four floors, totalling approximately 30,000 net usable square feet. There will be a series of large, open laboratories of 800 to 1,000 square feet. Each lab will accommodate six to ten investigators in animal, bacteria and plant tissue research. A number of smaller laboratories will be designated P-2 and P-3 for cultures of viable organisms containing recombinant DNA molecules. Conference room/library space will have the flexibility to handle groups of 75 to 100 people. Support facilities on each floor will include a darkroom, warm and cold rooms and an equipment room for high speed centrifuges and freezers. Space will be available for an electron microscope and computer graphics equipment. The facilities will be capable of supporting a total of approximately one hundred (100) people. On the roof of the building there will be a greenhouse for controlled propagation of plant material.

EXHIBIT C

Allocation of Indirect Costs

MGH may allocate to the Depatment indirect costs on the same basis as indirect costs are allocated to research grants and contracts awarded to MGH by the U.S. Department of Health and Human Services ("DHHS"). Indirect costs rates are negotiated periodically with the DHHS and the on-site rates in effect from time to time for such grants and contracts shall be the rates in effect under this Agreement. The base for such rates shall also be as determined for DHHS grants and contracts. Fringe benefit rates are also approved by DHHS. MGH shall apply the DHSS fringe benefit rates in effect from time to time to the Department.

For the period October 1, 1980 through September 30, 1981 the on-site indirect cost rate for DHHS grants and contracts is 50.71%, and the base is direct salaries and wages including vacation, holiday and sick pay but excluding other fringe benefits. Fringe benefit rates in effect from March 1, 1980 through September 30, 1980 were as follows:

Professional staff 15.2%

Nonprofessional staff 13.1%

THE MASSACHUSETTS GENERAL HOSPITAL

PARTICIPATION AGREEMENT

As a condition of my employment by The Massachusetts General Hospital (The Hospital) or my participation in any research (a) conducted by, under the auspices of, or pursuant to any agreement approved by The Hospital or (b) otherwise making substantial use of any facilities, materials, or other resources of The Hospital, and for other valuable consideration the receipt of which I hereby acknowledge, I hereby agree with The Hospital that, as to every discovery or invention (together hereinafter referred to as Invention) which, individually or jointly with others, I shall conceive or first reduce to practice during the course of such employment or participation:

1. I shall, in writing and reasonable detail, promptly disclose the Invention to the Director of The Hospital's Office of Technology Administration (Bartlett Hall, Third Floor; 726-2128);

2. In accordance with instructions from the said Director and at no expense to me, I shall execute and deliver such assignment of the Invention and other documents, and shall take all such other action pertaining to the Invention, as the Director may request of me in writing at any time once or oftener; and

3. I shall fully comply with The Hospital's Patent Policies and Procedures and Statement of Policy on Consulting Agreements (a copy of each of which is attached hereto and made a part hereof) and every obligation of The Hospital which shall apply to me under any grant or agreement providing support for research.

As a further condition of such employment or participation in such research, I also agree to file with the Director a complete and true copy of every agreement, if any, (1) to which I am party on the date hereof and (2) pursuant to which I am providing or shall provide consulting services which may materially relate to or draw on work (a) which I have done, am doing, or expect to do within the scope of my employment by The Hospital and (b) for which I have made, am making, or expect to make substantial use of facilities, materials, or other resources furnished by or through The Hospital.

Investigator:

________________________ ________________________________

Date

STATEMENT OF POLICY ON CONSULTING AGREEMENTS

The following Statement of Policy on Consulting Agreements has been adopted by the Trustees upon the recommendation of their Committee on Research and Committee on Patents. The Statement's purpose is (1) to protect the academic freedom traditional within The General Hospital, (2) to assist The Hospital's investigators (Investigator) and The Hospital itself in meeting their respective contractual responsibilities for research being carried on in normal course at The Hospital, and (3) to guide Investigators in evaluating invitations to provide compensated consulting services outside the scope of their employment by The Hospital or to accept additional compensation for services within or incidental to the scope of such employment (Consulting Agreement).

1. Such an invitation frequently offers the Investigator an opportunity to enlarge his* scientific knowledge and perspectives as well as additional remuneration. Thus, such invitations are welcome.

2. However, occasionally:

   a. The services contemplated by the proposed Consulting Agreement may materially relate to or draw on work (i) which the Investigator has done, is doing, or expects to do within the scope of his employment by The Hospital and (ii) for which he has made, is making, or expects to make substantial use of facilities, materials, or other resources furnished by or through The Hospital;

   b. Acceptance of the invitation may impose explicit or implicit restrictions on the freedom of the consulting Investigator (Consultant) to communicate with his colleagues at The Hospital about his consulting work, to publish reports on such work, to establish rights to own and use the fruits of the work, or to patent discoveries and inventions resulting from it -- i.e., restrictions on his entitlement to all or any of the foregoing academic freedom and intellectual property; and

   c. Such restrictions may impinge on rights and duties of both The Hospital and fellow Investigators of the Consultant at The Hospital, including obligations owed to sponsors of research already under way or contracted to be conducted at The Hospital.

---

*Includes the feminine gender throughout.

3. Thus, before accepting any proposed Consulting Agreement which shall require analysis of the possibility that any of the considerations noted in the foregoing Paragraph 2 may have application to him, every Investigator shall submit such proposal to the Director, Office of Technology Administration, (Director) and request that the Director (a) review the proposed agreement, (b) advise the Investigator as to its consonance with this Statement of Policy, and (c) if the Investigator shall so further request, assist him in negotiating and drafting appropriate revisions of the proposal. In so doing, the Investigator may delete from the document all financial terms specified therein.

4. Whenever, following such submission and review, the Investigator and the Director shall agree on the necessity of addressing any of the said considerations, the resulting Consulting Agreement, if any, executed by the Investigator as Consultant shall, subject to such exception or exceptions as the Consultant and the Director shall jointly determine to be appropriate, (a) incorporate by reference and be subject to:

i. Patent Policies and Procedures of The Massachusetts General Hospital in effect at the time being; and

ii. All additional obligations, if any, which the Consultant shall have at the time being under either or both of (aa) the patent policies of any institution other than The Hospital and (bb) any prior undertaking to conduct research, whether for The Hospital, pursuant to another Consulting Agreement, or otherwise; and

(b) impose no restriction on the freedom of the Consultant to discuss and disclose by publication or otherwise any research by him which shall make substantial use of any facilities, materials, or other resources furnished by or through The Hospital.

THE MASSACHUSETTS GENERAL HOSPITAL

PATENT POLICIES AND PROCEDURES

1.01 Basic Objectives. From time to time patentable discoveries or inventions are products of clinical, research, and educational activities, or any of them, undertaken under the authority and conventions of The Massachusetts General Hospital, excluding its McLean Hospital Division, ("The Hospital") by persons who are members of its Professional Staff or among its employees or students. As a non-profit institution striving to alleviate human suffering by engaging in the investigation, prevention, and treatment of disease, The Hospital's basic objectives in responding to and dealing with such discoveries and inventions are:

a. To promote their disclosure, dissemination, and utilization for the greatest possible public benefit;

b. To protect the rights to patents which their inventors and The Hospital may have; and

c. To provide for an equitable allocation of responsibilities and rewards among such inventors, The Hospital, and any organizations which may have sponsored and financed in whole or part any such activities ("Sponsor").

2.01 Patent Committee. The Committee on Patents ("Committee") shall consist of the General Director or his designee, the Director of Research Policy and Administration, the Chairman of the Committee on Research, a Trustee of The Hospital appointed for a term of three (3) years by the Chairman of The Hospital's Trustees ("Trustees' a member of the Corporation of The Hospital appointed for a term of three (3) years by the President of the Corporation, and, as non-voting ex officio members, the General Counsel or his designee and the Director of Fiscal Affairs or his designee. A quorum of the Committee shall be at least three (3) members eligible to vote; and, excepting adjournment, every action by the Committee shall require the presence of a quorum. The Committee's chairman shall be designated from time to time by the Chairman of the Trustees.

2.02 The powers and duties of the Committee shall be:

a. On behalf of The Trustees, to interpret and apply these Policies and Procedures ("Policies");

b. As of the close of each fiscal year of The Hospital, to report in writing to the Trustees on the activities of the Committee during such year, including the Committee's recommendations, if any, for amendment of these Policies; and

c. Such additional powers and duties as the Trustees may at any time once or oftener assign to the Committee.

3.01 Coverage. These Policies shall apply to every member of the Professional Staff, employee, and student of The Hospital who shall conceive or reduce to practice, actually or constructively, ("conceive") any discovery or invention while engaged in activities for which he or she shall receive financial support from The Hospital or during which he or she shall make substantial use of any facilities, materials, or other resources of The Hospital (every such person hereinafter referred to as "Inventor").

4.01 Patent Disclosure. Every Inventor shall, in writing and reasonable detail, give the Committee prompt notice of any discovery or invention which he or she shall desire to have patented or shall believe or have reason to believe is patentable.

5.01 Patent Ownership. The rights of ownership to every discovery or invention by any Inventor shall be the property of The Hospital; provided, however, that:

a. Within the one hundred and twenty (120) days next following disclosure under the preceding §4.01 by the Inventor of his or her discovery or invention (or such further period of time as may be agreed upon by the Inventor and the Committee), the Committee shall determine, and advise the Inventor in writing, whether such rights shall be retained by The Hospital or shall be released to the Inventor;

b. The rights of ownership to every discovery or invention conceived by any member of The Hospital's Professional Staff or any of its employees or students while engaged in activities for which he or she shall not receive financial support from The Hospital or during which he or she shall not make substantial use of any facilities, materials, or other resources of The Hospital shall be the property of such person;

c. The rights of ownership to every discovery or invention conceived during activities conducted pursuant to any agreement between The Hospital and any Sponsor shall be determined in accordance with such agreement; and

d. All disagreements over the rights of ownership to any discovery or invention not resolved by the Committee shall be referred to the Trustees for determination.

6.01 Seeking a Patent. Whenever the Committee shall determine to seek the patenting of any discovery or invention to which The Hospital shall have rights of ownership in whole or part, (a) The Hospital shall, without expense to the Inventor, provide such professional services as it shall deem to be necessary or desirable for such purpose and (b) the Inventor shall cooperate fully in such effort, including his or her execution of all necessary or desirable agreements, applications, and other forms and instruments. If at any time subsequently The Hospital shall terminate its effort to seek such patent, it shall promptly give written notice thereof to the Inventor, who thereupon shall be free at his or her expense to seek, develop, license, and otherwise use the patent.

7.01 Promotion and Licensing. In interpreting and applying these Policies, the Committee shall, by such means as it shall deem to be most effective and appropriate in each case, act to bring to the public all discoveries and inventions to which The Hospital shall have rights of ownership in whole or part. Such means may include, but shall not be limited to, agreements for the development, patenting, promotion, licensing, and manufacturing of any such discovery or invention; and in every case The Hospital shall advise the Inventor of the terms of any such agreement.

8.01 Proceeds from Patents. Subject to approval or modification by the Trustees in any case following their consideration of the written recommendation or recommendations of the Committee, the Net Proceeds or Annual Net Royalty (each as defined below in this section) received by The Hospital from any patent or unpatented invention or discovery owned in whole or part by The Hospital shall be apportioned and paid over by The Hospital in accordance with the following schedule:

Proceeds or Royalty ($)	Percentage (%) to Inventor	Inventor's Laboratory	Inventor's Department	Hospital's General Research Fund
First 50,000	35	35	15	15
Next 50,000 to 100,000	25	25	25	25
Over 100,000	15	15	35	35

provided, however, that the Committee may recommend in any case that more than one Inventor, laboratory, and department, or any of them, share in such apportionment: in which event (a) the payment apportioned under the said schedule to the Inventor shall be divided equally among all such Inventors and (b) the Committee shall specify each such laboratory and department, or either, and recommend the amount of payment to be made to each. Payment by The Hospital pursuant to any apportionment made to any Inventor in accordance with this §8.01 shall not be terminated for the reason that such Inventor shall cease to be a member of the Professional Staff, an employee, or a student of The Hospital. The terms "Net Proceeds" and "Annual Net Royalty," as used in and under these Policies, shall mean the net amount received by The Hospital in each fiscal year from the sale or licensing of any patent or unpatented invention or discovery owned in whole or part by The Hospital after deduction of all costs reasonably attributable to such patent, invention, or discovery, including without limitation any expense of patent prosecution and interference, litigation, and marketing.

9.01 Sponsors; Other Organizations. If and when any conflict shall arise between these Policies and any condition or conditions of (a) any proposed grant from or contract with any organization offering to act as a Sponsor or (b) the patent policies and procedures of any organization to which any Inventor shall have any obligation by virtue of any joint appointment or any affiliation or consulting agreement, such conflict shall be referred to the Committee. Following consideration of the conflict and such negotiations, if any, as the Committee shall deem to be warranted, the Committee shall, subject to the approval of the Trustees, determine and give appropriate notice of The Hospital's position in the matter.

10.01 Release of Rights of Ownership. Subject to the approval of the Trustees, the Committee may, for reasons and upon terms deemed to be satisfactory by it, release on behalf of The Hospital at any time any patent to its Inventor.

Appendix B
Washington University-Mallinckrodt, Inc., Agreement of 1981

HYBRIDOMA RESEARCH AGREEMENT

The Agreement is by and between Washington University (hereinafter "University"), a corporation organized under the laws of Missouri having its principal offices at Lindell and Skinker Boulevards, St. Louis, Missouri 63130 and Mallinckrodt, Inc. (hereinafter "Company"), a corporation organized under the laws of Missouri having its principal offices at 675 McDonnell Blvd., St. Louis, Missouri 63134;

WITNESSETH THAT:

WHEREAS, the University and the Company each have personnel and facilities for the conduct of research for the development of hybridomas that produce monoclonal antibodies;

WHEREAS, the University and the Company desire to expand research in the area of hybridomas, monoclonal antibodies and their applications;

WHEREAS, the University desires that the useful results of its research be made widely available to society through established avenues of trade and commerce;

WHEREAS, the Company has personnel and facilities for the research and development of medical products and production processes and for the manufacturing and wide distribution of medical products through established avenues of trade and commerce;

WHEREAS, the Company desires to add to its line of products advances utilizing hybridoma/monoclonal antibody technology;

WHEREAS, the parties hereto desire to cooperate and take advantage of each other's expertise for the research, development, production and marketing of useful products employing hybridoma/ monoclonal antibody technology, and

WHEREAS, the Company desires to sponsor research by the University aimed at the development of hybridomas that produce useful monoclonal antibodies and the University desires to have the Company sponsor such research;

NOW, THEREFORE, for and in consideration of the covenants herein contained, the parties do hereby agree as follows:

ARTICLE I - DEFINITIONS

1.1 "Program" shall mean (1) the research activities performed by the University directed to the development of hybridomas and monoclonal antibodies finding application in the areas of immunology and autoimmunity, lipoproteins and atherosclerosis, malignant disease, blood clotting factors and infectious diseases and (2) the research services for the production, screening and evaluation of such hybridomas and antibodies by the Core Laboratories (hereinafter defined) in support of said activities which research activities and services are approved and funded by the Advisory Committee from financial support provided by the Company.

1.2 "Project" shall mean a specific research activity within the Program, which research activity has been approved and funded by the Advisory Committee from financial support provided by the Company.

1.3 "Core Laboratories" shall mean the Central Hybridoma Production Laboratory, the Central Immunoassay Laboratory and the Diagnostic Evaluation Laboratory of the University's School of Medicine.

1.4 "Project Investigator" shall mean the scientist in charge of a Project and responsible for its conduct in accordance with the terms and conditions of the Project award and the established operating policies and procedures of the University. A Project Investigator shall be a faculty member qualified to be a principal investigator on research projects sponsored by the federal government and other governmental agencies.

1.5 "Technical Developments" shall mean any and all information, inventions, discoveries, advances, and know-how resulting from the Program, whether such information, discoveries, advances, and know-how are patentable or unpatentable.

1.6. "Patented Product" shall mean any product made, used or sold by Company or by any party acting on behalf of or under license from Company, which incorporates or is made by the use of any Technical Development made by University or jointly by University and Company which is furnished by University to Company pursuant to the terms of this Agreement and is covered by a claim of an unexpired patent or patent application obtained or filed pursuant to this Agreement in the country wherein the product is manufactured, used or sold, which claim has not been adjudicated invalid in a final adjudication from which there can no longer be an appeal. Patented Products shall not include any product covered by any claim of any such pending patent application if such application (1) has been finally rejected by the patent office or (2) is the first filed application in any country and has not issued within three years of its filing date or (3) is a continuing application of a previously filed application(s) and has not issued within three years of the filing date of the earliest filed application on which it is based.

1.7 "Proprietary Product" shall mean any product other than Patented Product which is made, used or sold by the Company or by any party acting on behalf of or under license from Company, which incorporates or is made by the use of any Technical Development made by University or jointly by University and Company which is furnished by University to Company pursuant to the terms of this Agreement. No product shall be a Proprietary Product if it is based on any such Technical Development which at the time of disclosure by University to Company (a) is part of the public domain through no fault of Company, or (b) was already known to Company.

1.8. "Net Selling Price" shall mean the gross invoice price received less trade discounts allowed and taken, allowance for returns, transportation charges included in such invoice price and sales and other excise taxes included in such invoice price.

ARTICLE II - PROGRAM RESPONSIBILITIES

2.1 Program activities in a subject area will consist of Projects selected by the Advisory Committee and carried out under the direction of Project Investigator(s).

2.2 Project Investigator(s) shall be responsible among other things for isolation of antigens for hybridoma production and for evaluating the specificity of the monoclonal antibodies produced.

2.3 The Central Hybridoma Production Laboratory responsibilities shall include:

a) production and sterility testing of tissue culture medium,

b) maintaining plasma cytoma and other cell lines for fusion,

c) preparation and cloning of spleen cell hybridomas,

d) cloning of hybridomas, e.g., soft agar,

e) maintaining clones in tissue culture and in frozen state and,

f) growing hybrids in mice as transplantable tumors.

2.4 The Central Immunoassay Laboratory responsibilities shall include:

a) screening the hybridoma culture media for immunoglobulin production,

b) screening the culture media for antibody activity to the immunogen,

c) identifying the class of immunoglobulins produced by hybridomas and monitoring preparations for continuity of performance and,

d) producing necessary reagents for analysis including antiimmunoglobulin and radioactive antigens.

2.5 The Diagnostic Evaluation Laboratory responsibilities shall include:

a) modifying and developing monoclonal antibody testing methods for practical utilization in the clinical laboratory setting,

b) making comparisons of modified and new methods to those currently in clinical laboratory use where feasible and,

c) testing of monoclonal antibodies for their potential as diagnostic reagents using clinical investigative protocols established in collaboration with Project Investigators.

2.6 After first informing the Advisory Committee, a Project Investigator may apply for and accept research support from other public and private agencies, including other commercial firms for research investigations which do not duplicate or diminish the commercial value of research being carried out under this Program, and such research support shall be excluded from this Agreement and shall not be administered as a part of this Program. Project Investigators shall not during the term of this Agreement seek research support from such agencies for research investigations within the subject areas of this Program, without first informing the Advisory Committee.

ARTICLE III - TERM

3.1 This Agreement shall be for a period of three (3) years commencing on the Effective Date which is September 1, 1981.

3.2 Within thirty (30) days after the end of the second year of this Agreement the parties shall enter into discussions as to whether both parties desire to continue the Program beyond the period of this Agreement. If the parties mutually determine that continuation beyond the said three (3) year period is desirable the parties shall proceed with negotiations to arrive at mutually acceptable terms and conditions for such continuation.

3.3 If, in accordance with Paragraph 3.2 the parties decide not to continue the Program beyond the stated period of this Agreement, then the Company shall have the right to elect to continue its support, on a Project by Project basis, for any Project started but not completed during the three year period. The Company shall make such elections and the parties shall negotiate in good faith mutually acceptable time extension and financial terms prior to the expiration of this Agreement. All other relevant terms of this Agreement shall apply to such terminal project continuations.

ARTICLE IV - PROGRAM ADMINISTRATION

4.1 The Program shall be under the control of the Program Advisory Committee consisting of five (5) members including Dr. Joseph M. Davie, Program Director, Dr. Paul Lacy, Dr. David Kipnis, Dr. Jay McDonald and Dr. Thomas Oesterling or another scientist appointed by the Company. Dr. Davie shall also be Chairman of the Advisory Committee. The Program Director shall be responsible for implementing the actions of the Advisory Committee and for the day-to-day administration of the Program.

4.2 Should the Program Director or any University member of the Advisory Committee be unable to continue service, replacements shall promptly be appointed by the University.

4.3 The Program Director shall convene a meeting of the Advisory Committee at least once each calendar quarter and otherwise as frequently as necessary to act on pending proposals, to review the financial status and progress of active Projects including work of the Core Laboratories, to deal with unanticipated problem areas, and to consider other matters concerned with the effectiveness of the Program. Except in an emergency, notice of a scheduled meeting shall be issued not less than one week prior to any such meeting. Any member may have any matter related to the conduct of the Program placed on the Advisory Committee agenda for the next meeting by making such a request in writing to the Program Director sufficiently in advance of the meeting to allow adequate preparation for a productive discussion of the matter.

4.4 The Program Director shall, after each meeting of the Advisory Committee, distribute to all Committee members, whether present at the meeting or not, a written summary of matters considered and actions taken.

4.5 Should the Company member of the Advisory Committee not be able to attend a meeting, the Company may designate an alternate representative by so notifying the Program Director on a meeting by meeting basis. However, it is understood by the parties that the effectiveness of the Advisory Committee will be promoted by continuity of membership and regular attendance at meetings by the members.

4.6 All actions to approve, defer or disapprove and to fund new Projects and basic activities of the Core Laboratories, to provide supplemental or continuation support to previously approved Projects or activities, and to discontinue previously approved Projects or activities shall be taken in convened meetings of the Advisory Committee. Any such action shall require approval of a majority of the members of the Advisory Committee.

4.7 All requests for Project grants and continuations and supplements thereof shall be on forms provided by the Program Director of the type attached hereto as Schedule 4.7. Such requests for new Projects shall include the name and C.V. of the Project Investigator, the requested start date, an estimated time to complete, the research aims, the importance of monoclonal antibodies to the target antigens, an outline of the experimental design and procedures, a detailed budget, other current, pending and planned sources of support for the Project Investigator, and agreement by the Project Investigator to conditions reflecting the provisions of this Agreement relating to assignment of proprietary rights, advance publication notice, distribution of cell lines, and acceptance of support for development of competitive products. The Program Director shall provide copies of Project Investigator requests to all members of the Advisory Committee at least fourteen (14) days prior to the meeting at which such requests are to be considered.

4.8 Award of Project grants, and revisions, continuations and supplements thereto, shall be in writing and shall incorporate such terms and conditions which the Advisory Committee deems necessary to assure performance of the Project and compliance with the provisions of this Agreement.

4.9 The Program Director shall obtain written annual progress reports from each Project Investigator, copies of which shall be provided to the Advisory Committee. The Program Director shall submit a written Program summary progress report to the Company upon the conclusion of each year of the term of this Agreement.

ARTICLE V - PROGRAM FINANCES

5.1 The Company hereby agrees to provide to the University for conduct of the Program during the term of this Agreement the total amount of three million, eight hundred eighty one thousand, two hundred fifty dollars ($3,881,250) which shall cover the expenses of the University.

5.2 The aggregate University spending plan for the Program over the term of this Agreement shall be as follows:

During the first year	- $1,293,750
Cumulative through the second year	- $2,587,500
Cumulative through the third year	- $3,881,250

The University shall use its best efforts to avoid exceeding the spending plan and shall request the Company's approval in advance in writing for any significant departure from said plan. The Company is not obligated to make any payment in excess of the total amount specified in paragraph 5.1 unless the parties mutually agree to increase said total amount by formal amendment of this Agreement. It is understood that funds not expended in a given year, except for the terminal year, shall be available for expenditure in succeeding years.

5.3 Payment by the Company to the University of the amount specified in paragraph 5.1 shall be in accordance with the following schedule:

Quarterly payments by check drawn to the order of Washington University and paid on or before the first day of each calendar quarter commencing on September 1, 1981 in the amounts of:

$215,625.00 for each of first 4 payments

$323,440.00 for each of next 8 payments

$107,807.50 for each of last 4 payments

However, at any time should cumulative expenditures by the University be significantly greater or less than cumulative payments hereunder due to accelerated spending approved by the Company under paragraph 5.2 or a lag in spending for any reason, then the University and the Company shall negotiate an adjustment in this payment schedule to more adequately reflect the actual and anticipated expenditures on the basis that payment of certain cost charges incurred during the first year are to be deferred until the year commencing on September 1, 1984.

5.4 Program funds granted by the Company to the University shall be used in the following Program areas:

Immunology and Autoimmunity area Projects	- $900,000
Lipoproteins and Atherosclerosis area Projects	- $450,000
Malignant Disease area Projects	- $450,000
Blood Clotting Factors area Projects	- $450,000
Infectious Diseases area Projects	- $450,000
Shared Costs of Core Laboratories	-$1,181,250
Total authorized funds	-$3,881,250

Any significant change in this allocation plan shall be approved by the Company in writing in advance.

5.5 All Program funds shall be administered by the Program Director who shall allot funds, with the approval of the Advisory Committee as specified in Article IV, to Projects and to the Core Laboratories. The Program Director shall monitor the Project and Core Laboratory's spending of allotted funds and may make adjustments among budget categories of approved budgets upon justified requests of Project Investigators and Core Laboratory directors. The Program Director shall keep the Advisory Committee informed of financial matters which might indicate a significant departure from Project or Core Laboratory plans previously approved by the Committee. The Program Director's financial records on all segments of the Program shall be available for review by any member of the Advisory Committee.

5.6 Approved funds for individual Projects and for support of individual Core Laboratories shall be maintained by the

University's Accounting Services in separate accounts for each such activity. Spending from each account shall be under the direct control of the cognizant Project Investigator and Core Laboratory director, respectively, who shall be furnished with the Accounting Services standard monthly statements of spending against their accounts.

5.7 The accounting records of Program activities shall be available for audit by the Company during the normal business hours of the University.

5.8 The University shall submit quarterly summary financial statements to the Company showing approved budgets and actual spending for each Project and Core Laboratory grouped by Program classes specified in paragraph 5.4 and compared with the total Program spending plan specified in paragraph 5.2.

5.9 The title to equipment purchased for use under this Agreement with Program funds shall vest in the University.

5.10 Upon termination of this Agreement the University shall provide a final accounting of Program funds to the Company within ninety (90) days following such termination. During said ninety (90) day period the University shall liquidate all outstanding obligations incurred prior to termination but shall not incur additional obligations. The balance of funds remaining shall thereupon be returned to the Company unless required for completion of Projects in accordance with paragraph 3.3.

ARTICLE VI - PUBLICATIONS AND CELL LINES

6.1 Program participants are at liberty to publish or disclose the results of their research, but the Company will be advised of the results before such results are disclosed to others outside of the University for purposes of protecting proprietary rights.

6.2 Through the mechanism set forth in paragraph 6.4 below, the Company shall seek to anticipate project results to minimize the need for delay of disclosure by promptly initiating actions to establish such rights, and to advise Program participants as early as possible of minimum practical precautions necessary to protect such proprietary rights. These precautions shall seek to minimize the material temporarily withheld from disclosure as well as the period of such temporary delay. Program grants will require the Program participants to provide copies of articles being submitted for publication to the Advisory Committee at least two (2) weeks before submission to the publishers for the purpose of screening for inventions on which patent applications have not been filed and for unauthorized disclosure of Company proprietary information. On written request by Company, University agrees to delay any such publication for up to three (3) months from the date of transmittal to the Advisory Committee to allow filing of applications or deletion of Company proprietary information.

6.3 The Company shall promptly review pre-publication articles to determine if potentially patentable inventions are disclosed and shall promptly thereafter inform the University of the Company's interest in obtaining patent rights to such inventions as provided for in Article VIII hereof.

6.4 The pre-publication reporting and evaluations as provided for in paragraphs 6.2 and 6.3 notwithstanding, the Company representative on the Advisory Committee is exposed to all Project plans before commencement of these Projects and such representative shall have the full opportunity and right to follow the progress of any and all Projects. Through this mechanism the Company shall determine as early as practicable the potential for establishing patent rights and its interest in obtaining a license of such rights. As soon as such potential is determined by the Company the parties shall cooperate on immediate actions necessary to the establishment of such rights. In this connection, the Project Investigators shall confer fully with Company regarding the performance of the Program hereunder, and shall make available for Company's inspection, at such reasonable times as the Project Investigators and Company determine all Technical Developments developed under this Agreement.

6.5 All publications reporting research results from Program activities shall acknowledge that support for such research was provided by the Company.

6.6 All Program awards will acknowledge that Program participants are free to distribute cell lines and/or products therefrom arising out of the Program to scientific colleagues employed by non-commercial organizations for research purposes which will not commercially benefit any third party. However,

Program awards will require that such distribution of such cell lines and/or products therefrom be under a written agreement of the type attached hereto as Schedule 6.6 between the University and the recipient scientific colleague and his employer. This Agreement shall require that such cell lines and/or products therefrom (1) be used for research purposes only, (2) not be used for commercial purposes or for the benefit of any commercial organization, and (3) not be made available to any other party. Program Awards will also require that the Company will be notified at least thirty (30) days prior to the proposed distribution to enable the Company to express its views on such proposed distribution, which views shall be duly considered. The University shall encourage Program participants to distribute antibodies of hybridoma cell lines and not the cell lines themselves when antibodies best serve the needs of the scientific colleague(s) requesting cell lines.

ARTICLE VII - CONFIDENTIALITY

7.1 The parties hereby acknowledge that University Technical Developments disclosed in requests for Project funding whether subsequently approved or not, Project progress reports whether written or oral, annual Program progress reports, discussions in Advisory Committee meetings and between Program participants and representatives of the Company, and in invention disclosures to the Company constitute valuable University property. Accordingly, the Company agrees for a period of ten (10) years from the date of receipt that it shall take reasonable precautions to safeguard in a manner comparable to that used to protect its own confidential technical information, not to disclose to others and to use only for the purpose of this Agreement University Technical Developments disclosed to the Company. However, the Company shall not be liable for unauthorized revelations or uses of University Technical Developments which occur in spite of such precautions. University Technical Developments to be so safeguarded shall be marked by the University as being proprietary, or if such University Technical Developments are conveyed by observation or conversation, the University shall inform the recipient of its proprietary nature. These obligations of confidence and non-use shall not extend to any University Technical Development which at the time of its disclosure by University to Company:

a. is part of the public domain or of the public knowledge through no fault of the Company; or

b. was in the possession of the Company, and such possession by the Company is documented prior to the date of such disclosure.

7.2. In order to implement the provisions of paragraph 7.1 above, the Company agrees that it shall restrict the dissemination of University Technical Developments received hereunder to only those persons whose knowledge of such University Technical Developments is reasonably necessary to the performance of the Company's obligations under this Agreement and it will advise such persons of the confidentiality and non-use requirements of this Agreement.

7.3 Since close cooperation between Company personnel and University personnel in the conduct of activities required by or contributing to the purposes of this Agreement may involve the disclosure of Company confidential information to such University personnel, it shall be the responsibility of the Company to obtain personal commitments of confidentiality as it deems necessary in the circumstances from University personnel requiring knowledge of Company confidential information.

ARTICLE VIII - PATENTS AND LICENSING

8.1 Based on data and information provided to the Company at the time of each Project award the Company shall within sixty (60) days thereafter provide to the University in writing a preliminary statement of its interest in commercializing the potential Technical Developments resulting from each Project.

8.2 In the case of any Project where the Company expresses positive interest in commercializing potential Technical Developments the Company shall monitor progress of the Project through its representative on the Advisory Committee or by such other arrangements as may be mutually acceptable to the Company and the Program Director. The primary purpose of such monitoring is to detect potentially patentable inventions as early as possible which shall be a responsibility of the Company.

8.3 When in the judgment of the Company potentially patentable inventions are first developed within a Project and/or Core Laboratories the Company shall make a written report of such to the University with its evaluation of the patentability and the potential importance of such inventions to commercialization prospects, and its election whether it wishes to pursue prospective patent rights to such inventions.

8.4 When in the judgment of the University a Project first reaches the stage where it has produced Technical Developments of apparent commercial utility and the University believes such Technical Developments may be patentable and have not yet been identified by the Company through the process described in paragraph 8.2, the University shall report such potentially patentable inventions to the Company. Thereupon, the Company shall make the evaluation and report to the University as specified in paragraph 8.3

8.5 When the Company has elected to pursue prospective patent rights to an invention it shall promptly cause its patent attorneys to file and prosecute in good faith a United States patent application on such invention. The Company shall also effect the filing and good faith prosecution of foreign patent applications corresponding to the United States application in whatever countries the Company elects to pursue prospective patent rights.

8.6 The Company agrees to bear the cost for filing and prosecution of patent applications under paragraph 8.5 and the issuance of patents thereon. The Company shall not be required to prosecute any such patent application beyond the point of final rejection by the United States Patent and Trademark Office or the equivalent stage of prosecution if a foreign application.

8.7 With respect to patent applications filed and prosecuted by the Company, the University, at its own expense, may designate and retain patent counsel of its own who shall be permitted to review such patent applications and proposed responses to Patent and Trademark Office actions thereon and to consult with the Company's patent attorneys before the filing thereof. However, the control of such filings and prosecutions shall rest with the Company.

8.8 Title to U.S. and foreign patent applications covering Technical Developments made by any person(s) acting on behalf of University filed under the provisions of paragraph 8.5, and any patent(s) subsequently issuing thereon shall be in University. Title to U.S. and foreign patent applications covering Technical Developments made by any persons acting jointly on behalf of University and Company filed under the provisions of paragraph 8.5 and any patent(s) subsequently issuing thereon shall be in University.

8.9 The parties shall do all acts necessary or reasonably desirable to provide the Company patent attorneys with all information and execution of all documents necessary or desirable in the evaluation of Technical Developments, and in the filing and prosecution of patent applications thereon, and in obtaining the issuance and maintenance of any patents issuing from such applications.

8.10 The University shall take all necessary and desirable actions to assure that it acquires sufficient title to all Technical Developments, patent applications and patents from those of its personnel making such so as to be entitled to grant the licenses specified in this Agreement to the Company.

8.11 The Company shall have the right of first refusal to obtain licenses from the University on Technical Developments resulting from the Program owned or controlled by the University in the sense of the University being able to grant licenses in such countries and in such fields of use as the Company may elect to take.

8.12 License grants to the Company pursuant to 8.11 of rights under patent applications and patents issuing thereon directed to Technical Developments made solely with Program support shall be exclusive for the life of such patents. For any Technical Development made with the joint support of the Program and funds provided by another sponsor, such license shall be exclusive for the life of the patents, if permitted by the other sponsor, and otherwise on the most favorable terms permitted, such terms to be not less than exclusive for a limited period of years or non-exclusive for the life of the patents.

8.13 License grants to the Company pursuant to 8.11 covering hybridoma cell lines and other Technical Developments, made with the support of the Program, on which no patent applications have been filed or are intended to be filed by either party, shall be exclusive for a period of not less than ten (10) years; provided, however, such licensed grants are subject to a right reserved in University to make and use hybridoma cell lines and other Technical Developments for research purposes only that will not commercially benefit any third party.

8.14 License grants to the Company shall contain the following terms and conditions:

a) requirement that the Company use reasonable efforts to bring about the lawful introduction of licensed products into the marketplace as early as is commercially practicable, consistent with the Company's usual and reasonable business practices and judgment,

b) requirement that the Company submit a product development plan specifying its best estimate of the schedule of key events to market entry and provide periodic reports of progress against this plan to the University,

c) right of the Company to sublicense others with prior knowledge of the University, such sublicensing arrangements to provide royalties to the University equivalent to what sublicensees would pay to the University if licensed directly on the same royalty terms as the Company,

d) a royalty schedule based on Net Selling Price with a scale of rates as follows:

(1) Patented Product, little or no direct, non-infringing product competition -- 8% - for life of patent

(2) Proprietary Product, with little or no direct competition -- 7% - for 7 years from the date of first sale

(3) Patented Product with significant direct, non-infringing competition -- 3% - for life of patent

(4) Proprietary Product with significant direct competition -- 2% - for 5 years from the date of first sale.

(5) Proprietary Product with significant direct competition, such competition resulting from a publication by University disclosing or leading directly to the competing product(s) -- 0%.

Rates (3), (4) and (5) applied on country by country basis require evidence of a sustained significant market share by competitive products in each country. Not earlier than (1) year after market entry the University agrees to negotiate a year by year reduction of rates (1) or (2) but not below rates (3) or (4) if the Company can demonstrate rates (1) or (2) are a significant contributing cause of a pricing disadvantage reflected in loss of market share,

(e) right of annual audit to confirm royalties on behalf of the University by firm to which the Company has no reasonable objection,

(f) indemnification of the University by the Company for liability arising from use of University Technical

Developments, and from use, sale or other disposition of products made by use of University Technical Developments, by the Company, sublicensees or any party acting on behalf of same. This provision shall survive termination of the license agreement,

(g) based on reasonable evidence that as a result of a distribution pursuant to paragraph 6.5 of a cell line and/or products therefrom resulting from the Program, a third party distributes a product which (1) significantly affects Company's business, and (2) incorporates or is made by the use of such cell line and/or products therefrom and such cell line and/or products therefrom are exclusively licensed to Company as a Proprietary Product hereunder, then Company's applicable royalties shall cease until such time as distribution of such product ceases or University brings suit to cause distribution of such product to cease. When such distribution ceases or suit is brought, payment of applicable royalties at the full rate shall resume. If requested by Company, University shall cooperate in investigations whether such cell lines and/or products therefrom originated with University,

(h) law of Missouri shall apply,

(i) such other provisions as the parties may mutually desire, and, in the case of an exclusive license of a jointly supported invention (see paragraph 8.12), such provisions as the other sponsor may require the University to include.

8.15 Should the Company not elect to license any particular Technical Development from the University, or elect to license and subsequently decide not to enter into the license agreement, or terminate the license agreement or should such agreement be terminated by the University free of a pending challenge by the Company, then the University shall be free to license such Technical Development to others. In the event University so licenses such Technical Development, Company shall receive 50% of the royalties received by University for such license after University's direct expenses for licensing such Technical Development have been deducted. Company shall continue to receive royalties for such license until it has been reimbursed for its funding of such Technical Development or it receives $25,000 whichever amount is greater. However, any such Technical Development shall not be directly competitive with and not substantially superior to any University Technical Development the Company has licensed and which is incorporated in scheduled product development plans which the Company is actively pursuing.

8.16 Upon the election by the Company to license University Technical Developments which are in the latter stage of research development under a Project, if the Company desires to commence activities directed at transferring such technology to its laboratories for commercial development, then the Program Director shall participate with the Company representative, the Project Investigator and such Core Laboratory directors as may be appropriate to work out mutually acceptable actions to be taken to effect such technology transfer.

ARTICLE IX - TERMINATION

9.1 This Agreement shall terminate on the expiration of the three (3) year period set forth in paragraph 3.1 hereof unless extended by mutual agreement of the parties as provided in paragraph 3.2 or 3.3.

9.2 In the event that either party to this Agreement defaults or breaches any of the provisions hereof, the other party reserves the right to terminate this Agreement upon ninety (90) days written notice to the defaulting party; provided that if the defaulting party, within said ninety (90) day period cures the said default or breach, this Agreement shall continue in full force and effect.

9.3 If either party shall become insolvent, or shall make any assignment for the benefit of creditors, or shall be adjudged bankrupt, or if a receiver or trustee of the property of either party is appointed, the other party on thirty (30) days written notice may terminate this Agreement.

9.4 Notwithstanding the termination of this Agreement for any reason the Company shall not thereby be relieved of the duties and obligations to make payments under the provisions of Article V and, if terminated prior to expiration of the term hereof, for University's obligations which cannot reasonably be cancelled and for accrued expenditures incurred before the effective date of termination. (Also see paragraph 5.10)

9.5 Notwithstanding the termination of this Agreement for any reason, the provisions relating to confidentiality and non-use set forth in this Agreement shall remain in effect.

9.6 If the University exercises its rights under paragraphs 9.2 or 9.3 and effects the termination of this Agreement, it shall be under no further obligation to grant further licenses to the Company and the Company shall promptly transfer to the University the prosecution of all pending patent applications not previously licensed to the Company.

9.7 Should this Agreement terminate by any means other than set forth in paragraph 9.6, the provisions of this Agreement providing for the election by the Company to license University Technical Developments made under the Program and for the filing and prosecution of patent applications thereon shall remain in full force and effect and the University shall negotiate required license agreements in good faith in accordance with the provisions of Article VIII hereof which shall also remain in effect until completion of negotiations of the last of such license agreements.

ARTICLE X - INDEMNIFICATION

10.1 The Company agrees to hold harmless, indemnify and defend the University from all liabilities, demands, damages, expenses and losses arising out of use by the Company, or by any party acting on behalf of or under authorization from the Company, of University Technical Developments or out of any use, sale or other disposition by the Company, or by any party acting on behalf of or under authorization from the Company, of products made by use of University Technical Developments. The provisions of this paragraph shall survive termination of this Agreement.

ARTICLE XI - TRANSFER OF INTEREST

11.1 Neither this Agreement, nor any of its rights and obligations, may be assigned, transferred or otherwise disposed of by either party without the prior written consent of the other unless such assignment, transfer or disposition is to a successor to all the business and assets of the transferor which pertain to the subject matter of this Agreement, and provided that such successor shall agree in writing with the other party to assume all the obligations of the transferor under this Agreement in a form satisfactory to the other party.

ARTICLE XII - NOTICE

12.1 Any notice or report required or permitted to be given under provisions of this Agreement shall be in writing and be mailed:

(a) If to the Company, to:

Dr. T. O. Oesterling
Vice President
Mallinckrodt, Inc.
675 McDonnell Blvd., P.O. Box 5840
St. Louis, MO 63134

with a copy to:

Mr. L. N. Goodwin
Corporate Patent Counsel
Mallinckrodt, Inc.
675 McDonnell Blvd., P.O. Box 5840
St. Louis, MO 63134

(b) If to the University, to:

Dr. Joseph M. Davie
Department of Microbiology and Immunology
Washington University School of Medicine
660 South Euclid Avenue
St. Louis, MO 63110

with a copy to:

Mr. Edward L. MacCordy
Associate Vice Chancellor for Research
Washington University
Lindell & Skinker Blvds.
St. Louis, MO 63130

12.2 In the event that either party shall change its address or the person(s) designated to receive notice, the party so changing shall notify the other in writing of the change.

ARTICLE XIII - GENERAL PROVISIONS

13.1 The Company shall not use the name of the University, its Medical School, its affiliated hospitals or its personnel in advertising or promotional materials pertaining to the subject matter of this Agreement without prior written consent of the University.

13.2 This Agreement shall be construed under the laws of the State of Missouri.

13.3 No waiver of any default, condition, provisions or breach of this Agreement shall be deemed to imply or constitute a waiver of any other like default, condition, provision or breach of this Agreement.

13.4 The Article headings used in this Agreement are for convenience only and form no part of the Agreement.

13.5 This writing constitutes the entire Agreement between the parties hereto relating to the subject matter of this Agreement and there are no understandings, representations or warranties of any kind except as expressly provided herein. Neither this Agreement, nor any term or provision thereof, may be discharged, waived, released, abandoned, changed or modified except by an instrument in writing signed by a duly authorized representative of each of the parties to this Agreement. If either party desires a modification or change of any kind to this Agreement, the parties shall, upon reasonable notice of the proposed modification or change by the party desiring the change, confer in good faith to determine the desirability of such modification or change.

13.6 The parties agree that it is the intention of neither party to violate any public policy, statutory or common laws, and governmental or supranational regulations; that if any sentence, paragraph, clause, or combination of the same is in violation of any applicable law or regulation, or unenforceable or void for any reason whatsoever, such sentence, paragraph, clause or combinations of the same shall be inoperative and the remainder of the Agreement shall remain binding upon the parties.

IN WITNESS WHEREOF, the parties have caused this Agreement to be executed in duplicate by their duly qualified officers.

Executed at St. Louis, Missouri, this 31st day of August, 1981.

MALLINCKRODT, INC.

By John L. Ufheil
John L. Ufheil
Executive Vice President

WITNESS:
Roy Joseph Klostermann

WASHINGTON UNIVERSITY

By Samuel B. Guze

Title Vice Chancellor for Medical Affairs

WITNESS:
Edward L. MacCordy

SCHEDULE 4.7

Project Investigator's Form

Conditions

In accepting this project support I agree to abide by the applicable terms and conditions of the contract between the University and the Company, which in summary are:

1. Proprietary rights to all inventions and discoveries (whether patentable or not) e.g., to hybridomas, including new methods related to their product or use, developed as a result, in whole or part, of this support will be assigned to Washington University for licensing in accordance with the support agreement with the sponsoring company. Investigators will cooperate in reporting and assisting in the establishment of such proprietary rights.
2. Medical school royalties resulting from the sale of products derived from hybridomas will be distributed as follows: 40% to the research laboratory(ies) in which the hybridomas were developed, 40% to the Department and 20% to the School of Medicine.
3. Publications: Scientific advances made during the progress of this research will be freely reported in the scientific literature. Each publication including abstracts should be submitted to the Advisory Committee two (2) weeks before it is submitted to the Journal. This will enable patent applications to be filed (if deemed appropriate) before publication occurs. Otherwise a brief delay of publication may be

necessary to allow filing necessary applications so as to avoid the loss of proprietary rights. Each publication will acknowledge support from this contract.

4. Distribution of hybridoma cell lines and products: Investigators at Washington University may freely distribute hybridoma antibodies produced under this contract to any other scientist they desire, who is employed by a non-commercial organization, for non-commercial research purposes only after the recipient signs an agreement with Washington University. Hybridoma cell lines, however, should be distributed only under exceptional circumstances and only after the recipient signs an agreement with Washington University to limit his use of the lines. Since patent applications (if any) must be filed before cell line distribution to avoid loss of patent rights, requests to distribute cell lines and/or antibodies (which are both proprietary property whether patented or not) should be submitted to the Advisory Committee one month before proposed distribution. Washington University retains sole control of commercial rights to both hybridoma cell lines and antibodies produced and will license such commercial rights in accordance with the agreement with the sponsoring company.

5. Other support: It is understood that Investigators receiving funds from this contract may, during the tenure of this

contract, apply for and accept research and contract funds from public and private agencies. Support from other sources is permitted during the tenure of this contract only to develop products that in no way duplicate or diminish the commercial value of products developed under this contract. Any additional support involving hybridoma research from any source should be reported to the Advisory Committee under item No. 10 of the Investigator's project request, or if occurring after this request, at the earliest time such proposal to another sponsor is planned or submitted.

6. Progress Reports: In order for the company to be fully informed about commercially important developments, several frequent brief progress reports may be requested in addition to a yearly report.

7. Collaboration with Company Scientists: When a commercially valuable reagent is produced, or at other times when an interaction is desirable between Company and University Scientists, the University Scientists may have access to Company confidential information; in this event the Company will require the University Scientist to sign a personal agreement indicating his willingness not to disclose such Company information to others.

It is understood that the Investigator in applying for a

grant has read and accepts the conditions and terms listed above.

Project Investigator:

Typed Name____________________

Signature_____________________

Date__________________________

Telephone_____________________

Core Laboratory Director's Form

Conditions

In accepting this project support I agree to abide by the applicable terms and conditions of the contract between the University and the Company, which in summary are:

1. Proprietary rights to all inventions and discoveries (whether patentable or not) e.g., to hybridomas, including new methods related to their product or use, developed as a result, in whole or part, of this support will be assigned to Washington University for licensing in accordance with the support agreement with the sponsoring company. Investigators will cooperate in reporting and assisting in the establishment of such proprietary rights.
2. Medical school royalties resulting from the sale of products derived from hybridomas will be distributed as follows: 40% to the research laboratory(ies) in which the hybridomas were developed, 40% to the Department and 20% to the School of Medicine.
3. Publications: Scientific advances made during the progress of this research will be freely reported in the scientific literature. Each publication including abstracts should be submitted to the Advisory Committee two (2) weeks before it is submitted to the Journal. This will enable patent applications to be filed (if deemed appropriate) before publication occurs. Otherwise a brief delay of publication may be

necessary to allow filing necessary applications so as to avoid the loss of proprietary rights. Each publication will acknowledge support from this contract.

4. Distribution of hybridoma cell lines and products: Investigators at Washington University may freely distribute hybridoma antibodies produced under this contract to any other scientist they desire, who is employed by a non-commercial organization, for non-commercial research purposes only after the recipient signs an agreement with Washington University. Hybridoma cell lines, however, should be distributed only under exceptional circumstances and only after the recipient signs an agreement with Washington University to limit his use of the lines. Since patent applications (if any) must be filed before cell line distribution to avoid loss of patent rights, requests to distribute cell lines and/or antibodies (which are both proprietary property whether patented or not) should be submitted to the Advisory Committee one month before proposed distribution. Washington University retains sole control of commercial rights to both hybridoma cell lines and antibodies produced and will license such commercial rights in accordance with the agreement with the sponsoring company.

5. Other support: It is understood that Investigators receiving funds from this contract may, during the tenure of this

contract, apply for and accept research and contract funds from public and private agencies. Support from other sources is permitted during the tenure of this contract only to develop products that in no way duplicate or diminish the commercial value of products developed under this contract. Any additional support involving hybridoma research from any source should be reported to the Advisory Committee under item No. 10 of the Investigator's project request, or if occurring after this request, at the earliest time such proposal to another sponsor is planned or submitted.

6. Progress Reports: In order for the company to be fully informed about commercially important developments, several frequent brief progress reports may be requested in addition to a yearly report.

7. Collaboration with Company Scientists: When a commercially valuable reagent is produced, or at other times when an interaction is desirable between Company and University Scientists, the University Scientists may have access to Company confidential information; in this event the Company will require the University Scientist to sign a personal agreement indicating his willingness not to disclose such Company information to others.

It is understood that the above terms, applicable to each Project Investigator, apply equally and are acceptable to the Core Laboratory Director whose laboratory receives support under the University's contract with the Company.

Core Laboratory Director:

Typed Name____________________

Signature_____________________

Date__________________________

Telephone_____________________

SCHEDULE 6.6

WASHINGTON UNIVERSITY

Ref: ______________

AGREEMENT BETWEEN WASHINGTON UNIVERSITY AND RECIPIENT SCIENTIST AND INSTITUTION FOR DISTRIBUTION OF CELL LINE AND/OR PRODUCTS THEREFROM.

1. Under the terms of this Agreement the following cell line and/or products therefrom (hereafter "material") are to be made available:

____

____

2. The Recipient Scientist and Institution agree to use this Material solely for application in the non-commercial research of the Recipient Scientist at the Institution.

3. This Material will not be distributed by the Recipient Scientist or the Institution to any other person or organization, except those scientists affiliated with and working within the Institution in collaboration with the Recipient Scientist who have also accepted these terms by signing below. Distribution to any other scientist of the Institution shall be the subject of a separate agreement with Washington University.

4. This Material is the property of Washington University and is provided as a service to the academic research community. Its distribution is limited to scientists at non-profit institutions solely for use in their own research. This Material may not be used by the Recipient Scientist or the Institution for commercial purposes nor for the benefit of any for-profit organization. Should unauthorized organizations obtain this Material, upon request by Washington University the Recipient Scientist and Institution agree to advise Washington University of their knowledge, if any, of such Material transfer to the unauthorized organization.

5. This Material is experimental in nature and its characteristics are not completely known. The Recipient Scientist and Institution agree to use special care in the handling, storage, use and disposal of this Material commensurate with guidelines published by the National Institutes of Health.

6. Acknowledgement of the source of this material will be made in relevant publications in the following form:

____

____

____

7. The Recipient Scientist and Institution warrant that they are entitled to receive this Material under all applicable laws and regulations. This Material is provided without warranty of merchantability or fitness for a particular purpose or any other warranty, expressed or implied. The Recipient Scientist and Institution agree to waive all claim against Washington University, and to defend and indemnify Washington University from all claims and damages asserted by third parties arising from actions by the Recipient Scientist and/or the Institution and its employees.

8. If the above terms are acceptable, please have the Recipient Scientist, an authorized official of the Institution and collaborating scientists (if any) sign in the spaces below. Return the signed Agreement to:

Chairman
Department of Microbiology & Immunology
Washington University School of Medicine
660 South Euclid Avenue
St. Louis, MO 63110

The undersigned agree to the foregoing:

Recipient Scientist

____

Collaborating Scientist(s)

____

____

____

____

For the Institution:

____
(authorized official)

____
(Title)

____
(Institution)

Date: ____________________

Appendix C
Washington University-Monsanto Biomedical Research Agreement of 1982

This Agreement, effective as of July 1, 1982, is by and between the parties:

WASHINGTON UNIVERSITY, a corporation organized under the laws of Missouri and having its principal offices at Lindell and Skinker Boulevards, St. Louis, Missouri 63130 (hereinafter "University")

AND

MONSANTO COMPANY, a corporation organized under the laws of Delaware and having its principal offices at 800 North Lindbergh Boulevard, St. Louis, Missouri 63167 (hereinafter "Monsanto");

WITNESSETH THAT;

WHEREAS, the University has sought and continues to seek the advancement of knowledge through education and research;

WHEREAS, the University desires that the useful results of its research be made available to society through established avenues of trade and commerce;

WHEREAS, Monsanto has personnel and facilities for the conduct of research, for the development of new products and processes based on scientific research, and for efficient large scale manufacture and distribution;

WHEREAS, Monsanto seeks to utilize the fruits of scientific research as a source for the development, manufacture and distribution of new products, especially products for meeting human needs;

WHEREAS, the University and Monsanto recognize that each can benefit from a relationship in biomedical research extending over a span of years that will provide present and potential financial support for the University, potential benefit to health care consumers and potential commercial benefit for Monsanto, while enhancing the understanding and work of their respective scientists by close interaction among them;

WHEREAS, the University and Monsanto believe that industrial support of biomedical research can lead to enhancement of their respective capabilities and render important long range benefits to the University, to Monsanto and to society;

WHEREAS, the University and Monsanto believe that biomedical inventions are likely to be brought into public use for public benefit through the incentive of the protection of the Patent System utilized by the parties to make available through Monsanto, new commercial products and processes, while concurrently providing royalty income to the University to support its educational and charitable activities;

WHEREAS, the University and Monsanto recognize that the concept of academic freedom must be preserved by this Agreement and shall be a guiding principle in its administration;

WHEREAS, the University and Monsanto recognize that the 1964 Statement on Preventing Conflicts of Interest in Government Sponsored Research at Universities, issued by the American Association of University Professors and the American Council on Education expresses principles applicable to corporate and university relationships;

WHEREAS, the University and Monsanto are prepared to undertake a collaborative effort in the field of biomedicine with a focus on proteins and peptides which modulate cellular function, where the University currently has substantial personnel and facilities for the conduct of research and a field where Monsanto has in-house research underway and wherein Monsanto expects to increase its in-house research emphasis; and

WHEREAS, Monsanto proposes to provide significant financial support to the University in furtherance of this collaborative effort according to the terms set forth in this Agreement.

NOW, THEREFORE, the parties hereto agree as follows:

ARTICLE I

PURPOSE AND SCOPE OF THIS AGREEMENT

The purpose of the present Agreement is to provide a contractual framework to govern conduct of this collaborative effort under which multiple research Projects (as hereinafter defined) can be undertaken. This Agreement is designed to recite the contractual provisions which would apply to all Projects authorized by the Advisory Committee under the Program (as hereinafter defined).

ARTICLE II - DEFINITIONS

2.1 "Program" means all research activities performed by or for the University under this Agreement which are authorized and funded by the Advisory Committee (as hereinafter defined) and Program Director from financial support provided by Monsanto.

2.2 "Project" means a specific research activity which has been authorized and funded by the Advisory Committee from financial support provided by Monsanto under the Program. Projects shall be of three types:

a) "Exploratory Projects": Those directed to fundamental research on basic scientific questions with a focus on proteins and peptides which modulate cellular function.

b) "Specialty Projects": Those directed to applied research with a focus on proteins and peptides which modulate cellular function and in which Monsanto sees more immediate commercial utility either in terms of technologies or products or both.

c) "Construction and Renovation Projects": Those construction and renovation activities directed to physical facilities required to accommodate and enhance the Program.

2.3 "Advisory Committee" means those representatives of the University and Monsanto charged with administering the Program. The Advisory Committee comprises a Program Director who shall be Chairman and appointed by the University, three (3) additional members appointed by the University, and four (4) members appointed by Monsanto. All members including the Program Director, shall have voting power.

2.4 "Project Investigator" means the scientist in charge of a Project and responsible for its conduct in accordance with the terms of the Project award and the accepted operating policies and procedures of the University. A Project Investigator shall be a faculty member qualified to be a principal investigator on research projects sponsored by government and nationally reputable agencies.

2.5 "Technical Developments" means any and all inventions, discoveries, advances, know-how, processes, devices, machines, materials, software and other information arising from the Program, whether or not the same are patentable, copyrightable or otherwise protectable by law.

2.6 "Patent" means any patent, certificate of invention, inventors certificate, utility model or similar form of protection, or plant patent or other form of protection of plant material, granted anywhere in the world covering an invention which is a Technical Development, and owned by the University or in which the University has licensing rights.

2.7 "Licensed Product" means any product covered by a claim or made by or used in a process covered by a claim of an unexpired Patent at the time and in the country wherein the product is manufactured, used or sold, which claim has not been adjudicated invalid in a final adjudication from which there can no longer be an appeal, and which Patent is licensed to Monsanto as provided for in this Agreement.

2.8 "Agreement of Program Participants" means the specimen agreement set forth in Exhibit A attached hereto.

ARTICLE III - TERM OF AGREEMENT

3.1. This Agreement shall be for a period of five (5) years commencing July 1, 1982 and terminating June 30, 1987, unless earlier terminated under the provisions of Paragraphs 4.3, 12.2 or 12.3.

3.2 On or about February 1, 1985, the parties shall enter into discussions as to whether both parties desire to continue the Program beyond the normal termination date of June 30, 1987. If continuation is mutually desirable the parties shall proceed with negotiations to arrive at mutually acceptable terms and conditions for such continuation. If continuation is not desired by either or both parties, this fact shall be confirmed in writing before the end of the third year of the initial term of this Agreement.

3.3 If, in accordance with Paragraph 3.2 the parties decide not to continue the Program beyond June 30, 1987, then Monsanto shall have the option of electing to continue its support, on a Project by Project basis, for any Project started but not completed during the normal term. Monsanto shall make such elections and the parties shall negotiate in good faith mutually acceptable financial terms and time extensions, not to exceed two (2) years in duration, prior to the expiration of this Agreement. All other relevant terms of this Agreement shall apply to such terminal Project continuations.

ARTICLE IV - PROGRAM ADMINISTRATION

4.1 The Program shall be under the direction of the Advisory Committee chaired by the Program Director, Dr. David M. Kipnis, who shall be assisted by seven (7) other Committee members including three (3) members, namely, Dr. Luis Glaser, Dr. Paul Lacy, and Dr. Joseph Davie, appointed by the University and four (4) members, namely, Dr. Howard A. Schneiderman, Dr. G. Edward Paget, Dr. Louis Fernandez and Dr. David C. Tiemeier, appointed by Monsanto. The University and Monsanto representatives on the Advisory Committee, other than the Program Director, may be changed at appropriate intervals by either of the parties with timely notice to the other party.

4.2 All actions to approve, defer or disapprove Program activities and to fund new Projects, to provide supplemental or continuation support to previously approved Projects or activities, and to discontinue previously approved Projects or activities shall be taken in convened meetings of the Advisory Committee. Any such action shall require approval of a majority of the members of the Advisory Committee, i.e., at least five (5) of the eight (8) members.

4.3 Should the Program Director or any member of the Advisory Committee be unable to continue service, a replacement shall be promptly appointed by the appropriate party. Program Director replacements shall be mutually acceptable to Monsanto and the University; provided, however, that acceptance by Monsanto shall not be unreasonably withheld. If the University cannot nominate an acceptable replacement for the Program Director within one (1) month following the inability of the Program Director to continue service, Monsanto may suspend its financial support for the Program until an acceptable Program Director is appointed. If such suspension continues beyond six (6) months, Monsanto may summarily treat this Agreement as breached under provisions of Paragraph 12.2 and the ninety (90) day notice provision of Paragraph 12.2 is not applicable.

4.4 The Program Director shall convene a meeting of the Advisory Committee at least once each calendar quarter and otherwise as frequently as necessary to act on Program matters and pending proposals, to review the financial status and progress of active Projects, to deal with unanticipated problem areas, and to consider other matters concerned with the effectiveness of the Program. Except in an emergency, notice of a scheduled meeting and an agenda therefor shall be issued not less than two (2) weeks prior to any such meeting. Any Advisory Committee member may request convening of special meetings and may have any matter related to the conduct of the Program placed on the Advisory Committee agenda for the next or forthcoming meeting by making such a request in writing to the Program Director sufficiently in advance of the meeting to allow adequate preparation for a productive discussion of the matter.

4.5 The Program Director shall, after each meeting of the Advisory Committee, distribute to all Committee members, whether present at the meeting or not, a written summary of matters considered and actions taken.

4.6 Should a member of the Advisory Committee not be able to attend a given meeting, an alternate representative may be designated by so notifying the Program Director on a meeting by meeting basis. If the Program Director is unable to attend a meeting of the Advisory Committee, he may designate another University member of the Advisory Committee to chair the meeting and perform the functions of the Program Director at that meeting. However, it is understood by the parties that the effectiveness of the Advisory Committee will be promoted by continuity of membership and regular attendance at meetings by members.

ARTICLE V - PROJECT SELECTION AND IMPLEMENTATION

5.1 The Advisory Committee shall decide on both the Exploratory and Specialty Projects which are to be supported under the Program. The Advisory Committee shall strive to identify and fund Projects in which the University enjoys scientific leadership and in which Monsanto has a meaningful interest.

5.2 The Advisory Committee has ultimate responsibility for identification and selection of all Projects as well as for overall and ongoing direction of the Program. As a general guide, the parties to this Agreement intend for the Program to embrace two (2) types of Projects, namely, Exploratory Projects and Specialty Projects. Ultimately during the term of this Agreement, it is expected that approximately thirty percent (30%) of the research effort would be directed toward fundamental questions (Exploratory Projects) while seventy percent (70%) would be directed toward specific products (Specialty Projects). The parties hereto recognize that facility renovation and construction is to be funded as a Program activity within the limitation of the financial support specified in Article VIII hereof.

5.3 Following the identification of a field of interest by the Advisory Committee the Program Director shall seek Project proposals from faculty members of the University.

5.4 Project proposals, continuations and supplements thereto shall be on forms provided by the Program Director. The Program Director shall provide copies of Project proposals to all members of the Advisory Committee at least one (1) month prior to the Committee meeting at which such requests are to be considered.

5.5 Whenever the Advisory Committee has identified a field of research of mutual interest, and has received an acceptable Project proposal, a Project may be created by the authorization of the Advisory Committee in writing. The Project authorization shall identify the Project Investigator, define the research activities to be pursued, the level of effort to be devoted to the Project by the Project Investigator, include a budget covering all costs of such research, define the time duration and such other terms and conditions as may be agreed to and be approved by the Project Investigator consistent with the purposes and conditions of this Agreement.

5.6 With concurrence of the Advisory Committee, and in furtherance of productive interaction between scientists of Monsanto and those of the University, Monsanto representatives on the Committee shall designate a Monsanto Project Scientist who shall act as the primary contact with each Project Investigator during the conduct of a given Project.

5.7 The Program Director shall submit to Monsanto in writing summary reports of all important findings and results as soon as available and detailed annual Program reports on each anniversary of this Agreement. The annual reports shall include summaries and conclusions for each active Project.

ARTICLE VI

INTERACTION BETWEEN MONSANTO AND THE UNIVERSITY

6.1 To optimize the mutual benefit and collaboration intended by this Program, the parties desire that there be mutually productive and continuing interchanges between University and Monsanto scientists. Accordingly, the University will ensure that all University scientists engaged in the Program are available to appropriate Monsanto scientists for consultation in the area of their respective Projects. Temporary office space at the University shall be made available to collaborating Monsanto scientists.

6.2 The University agrees to permit individual scientists and technicians from Monsanto, with the consent of the Program Director and Project Investigator and at Monsanto's expense, to spend appropriate periods of time in University laboratories where Project research is being conducted to learn techniques developed therein, to participate if mutually desirable, and to facilitate the transfer of Technical Developments to Monsanto. Monsanto agrees that its employees who are permitted to train and function in the laboratories of the University pursuant to this paragraph shall be required to observe the applicable policies of the University.

6.3 It is anticipated that interaction between the Project Investigators and Monsanto Project Scientists will identify facilities and capabilities of Monsanto which may be used by University scientists to enhance the progress of Projects. Moreover, it is appropriate that evaluation of the commercial potential of research leads and products be addressed through the interaction of the Project Investigators and the Monsanto Project Scientists.

ARTICLE VII - SCIENTIFIC REVIEW PANEL

7.1 To assess the scientific merit and cost effectiveness of Projects supported by the Program, the parties hereto recognize the need for periodic review by an independent panel of scientists.

7.2 During the third year of the initial term of this Agreement and every two (2) years thereafter, the Advisory Committee shall commission a scientific review panel comprising at least four (4) distinguished scientists, not employees of Monsanto or members of the University staff, to review all then-current Project work and to appraise the direction of the Program, both qualitatively and quantitatively. Composition of the review panel should be designed to include scientists having clinical and pharmaceutical orientation as well as academic orientation.

7.3 The review panel shall be required to issue a confidential report to the Advisory Committee and to the Chancellor of the University and the Chief Executive Officer of Monsanto stating its views, conclusions and recommendations regarding the scientific merit and cost effectiveness of the Program and Projects and the impact of the Program on the respective institutions involved.

7.4 Costs of the scientific review shall be paid from Program funds.

VIII - PROGRAM FINANCES

8.1 Monsanto hereby agrees to provide to the University for the total support of the Program during the five (5) year term of this Agreement, the total amount of Twenty-Three Million, Five Hundred Thousand Dollars ($23,500,000), to be adjusted according to Paragraph 8.2, which shall cover both direct and indirect expenses of the University. The University agrees that this funding shall be disbursed solely in support of the Program.

8.2 Payment by Monsanto to the University of the amount specified in Paragraph 8.1 shall be limited to contract year budget amounts recited in the following schedule which are subject to (i) annual adjustment for inflation in accordance with this Paragraph 8.2, and (ii) budget underruns carried forward from one year to the next with approval of the Advisory Committee in accordance with Paragraph 8.9. The parties hereto believe the following expenditure schedule reflects the appropriate allocation of funds:

Contract Year	Exploratory Projects	Specialty Projects	Construction and Renovation Projects	Contract Year Budget
82/83	$ 1,500,000	$ 1,500,000	$ (See Para.8.4)	$ 3,000,000
83/84	$ 1,600,000	$ 2,200,000	$	$ 3,800,000
84/85	$ 1,700,000	$ 3,000,000	$	$ 4,700,000
85/86	$ 1,800,000	$ 3,800,000	$	$ 5,600,000
86/87	$ 1,900,000	$ 4,500,000	$	$ 6,400,000
Total	$ 8,500,000	$ 15,000,000	$	$ 23,500,000

The initial contract year shall run from the effective date of this Agreement through June 30, 1983. Subsequent contract years shall run from July 1 through June 30.

The contract year budgets above recited, commencing with the second contract year (July 1, 1983 through June 30, 1984), shall be adjusted using the GNP Deflator Index in the following manner:

(a) A base index will consist of an average of the GNP Deflator Index figures for the four

(4) quarters from April 1981 through March 1982.

(b) An index for each contract year, commencing with the second contract year, will consist of an average of the four (4) quarterly GNP Deflator Index figures covering the period April through the following March immediately preceding the start of each contract year. (For example the index for the second contract year will be the average of the GNP Deflator Index figures for the four (4) quarters covering April 1982 through March 1983.)

(c) Each contract year budget as stated above shall be adjusted prior to the commencement of the relevant contract year by applying a multiplier derived as follows:

$$\text{multiplier} = 1 + \frac{\text{contract yr. index} - \text{base index}}{\text{base index}}$$

For purposes of this Agreement the "GNP Deflator Index" shall mean the quarterly revised Implicit Price Deflator for the Gross National Product as reported by The United States Department of Commerce, Bureau of Economic Analysis. Since it is normal for a quarterly GNP Deflator Index to be revised shortly after it is first published, calculations herein shall be based on the final index for a quarter, if available, and otherwise on the most recent revision available on June 1 immediately preceding the start of the contract year for which calculations are made.

8.3 It is recognized that the occurrence of expenditures during a contract year is primarily dependent on Project spending plans authorized by the Advisory Committee during the current and any prior years. Nevertheless, Monsanto is not obligated to reimburse the University for expenditures incurred during, or carried forward into, any contract year in excess of the total amount of the contract year budget shown on the expenditure schedule in Paragraph 8.2, as it may have been adjusted under the provisions of Paragraph 8.2 and 8.9, unless the parties mutually agree to modify said total amount by formal amendment to this Agreement.

8.4 All Program funds shall be administered by the Program Director who shall allot funds, with the approval of the Advisory Committee as specified in Article IV, to Project participants. By unanimous consent the Advisory Committee may reallocate among Project types up to 10% of the total funds for any contract year specified in the schedule of Paragraph 8.2, as such annual total may have previously been modified by Monsanto under Paragraph 8.3 or by the Advisory Committee under Paragraph 8.9. Such reallocation of contract year funds may be among the

Exploratory Project type, the Specialty Project type and the Construction and Renovation Project type. The Program Director shall monitor spending of funds budgeted for individual Projects and may make adjustments among expense categories of an approved Project budget upon justified requests of Project Investigators. The Program Director shall keep the Advisory Committee informed of financial matters which might indicate a significant departure from Project plans previously approved by the Committee. The Program Director's financial records on all segments of the Program and Projects shall be available for review by any member of the Advisory Committee.

8.5 Approved funds for individual Projects or for support of the Program shall be maintained by the University's Accounting Services Department in separate accounts for each such activity. Spending for each account shall be under the direct control of the Program Director or his delegated Project Investigator, respectively, who shall be furnished with the Accounting Services standard monthly statements of spending against their accounts.

8.6 The accounting records of Program activity shall be available for audit by Monsanto, using its own internal or outside auditors, during the normal business hours of the University.

8.7 The University shall submit monthly invoices with

supporting details to Monsanto showing actual spending by University expense category for each Project for which reimbursement of expenditures is being requested. Each invoice shall also show cumulative expenditures to date for each such Project against the approved Project budget and cumulative total Program expenditures for the contract year against the current contract year budget shown on the expenditure schedule in Paragraph 8.2 as it may have been previously adjusted under the provisions of Paragraphs 8.2 and 8.9.

8.8 Monsanto agrees to pay the University promptly upon receipt and approval of the University's invoices provided under Paragraph 8.7 up to the level of the contract year budget set forth in Paragraph 8.2, as such contract year budget may have been adjusted under the provisions of Paragraphs 8.2 and 8.9.

8.9 If in any contract year there is an overrun of the contract year budget the excess expenditures shall be carried forward and be paid from the following contract year budget. If in any contract year there is an underrun of the contract year budget (hereinafter in this paragraph "the current contract year budget"), then with the unanimous consent of the Advisory Committee the underrun amount may be carried over as an addition to the following contract year budget. The approved amount from the current contract year budget which is to be carried over shall be adjusted by a multiplier calculated by dividing the multiplier from Paragraph 8.2 for the following contract year budget by the multiplier for the current contract year budget. The thus adjusted amount to be carried over shall then be added to the following contract year budget after the following contract year budget has been adjusted in the usual manner.

8.10 Title to all items of equipment purchased with Program funds shall vest in the University at the time of purchase.

8.11 Upon termination of this Agreement for any reason the University shall provide a final accounting of Program funds to Monsanto within ninety (90) days following such termination. During said ninety (90) days the University shall liquidate all outstanding obligations incurred prior to termination but shall not incur additional obligations. The balance of funds remaining shall thereupon be returned to Monsanto unless required for completion of Projects in accordance with Paragraph 3.3.

8.12 Indirect costs invoiced under Paragraph 8.7 shall, through June 30, 1987, be at a fixed rate of fifty percent (50%) of invoiced direct costs. Indirect costs invoiced by the University for any activity performed in whole or in part by any contractor shall not exceed the indirect costs which would have been invoiced had such activity been performed wholly by the University. If the University's indirect costs rise by ten percent (10%), i.e., to fifty five percent (55%) or more, then upon the University's request Monsanto agrees that it will negotiate the University's request to increase the rate of indirect costs from fifty percent (50%) under this Agreement, taking into consideration relevant factors, including relative increases in indirect costs made in other research agreements, including government agreements.

ARTICLE IX - PUBLICATIONS AND REVIEW OF TECHNICAL DEVELOPMENTS

9.1 The University faculty members participating in Projects are at liberty to publish the results of their research subject to the provisions of Paragraphs 9.1, 9.2, 9.3, 9.4 and 9.5. Project awards will require that participants provide copies of all abstracts and articles, in the best form then available, proposed to be submitted for publication in sufficient time to permit the Program Director to provide same to a Monsanto member of the Advisory Committee at least one (1) month prior to submission to a publisher or other third party. The Program Director shall immediately determine that a Monsanto member has received a copy of each such proposed abstract and article. The Program Director shall also promptly provide to a Monsanto member a final copy of each abstract and article as submitted for publication.

9.2 Monsanto shall promptly review such proposed abstracts and articles to determine if potentially patentable Technical Developments are disclosed and shall promptly thereafter inform the University whether delay of submission for publication or other public disclosure for a reasonable time will be required to establish Patent rights of reasonable scope. Disputes concerning such delays shall be referred to the Advisory

Committee.

9.3 As to verbal presentations and discussions, the parties recognize that it is impractical to provide a complete review system for Patent purposes and that considerable discretion must be left in the investigator. It is the intent of the University and Monsanto to provide the investigators guidance sufficient to avoid any divulgations that would compromise the establishment of the best possible Patent position.

9.4 The reporting and evaluation as provided for in Paragraphs 9.1 and 9.2 notwithstanding, the Monsanto representatives on the Advisory Committee are exposed to all Program and Project plans before commencement and such representatives have full opportunity and right to follow the progress of any and all Projects. Through this mechanism the assigned Monsanto Project Scientists and Monsanto shall determine as early as practicable the potential for establishing Patent rights and its interest in obtaining a license of such rights. As soon as such potential is determined by Monsanto the parties shall cooperate on immediate actions necessary to the establishment of such rights, including, if necessary delay of publication for a reasonably brief period of time to conduct any further research or take other actions that may be necessary to file appropriate and adequate Patent applications.

9.5 All scientific publications reporting research

results from Program activities shall acknowledge that support for such research was provided by Monsanto.

9.6 Upon written request to the Advisory Committee, Monsanto shall receive adequate samples of all available scientific materials isolated or developed in the Program, and shall have the right to use the same for research and/or commercial purposes, but subject to the provisions herein with respect to confidentiality, Patents and licenses. Monsanto's rights to receive and use samples as provided in this Paragraph 9.6 shall not be denied but shall be subject to reasonable modification for good reason as deemed necessary by the Advisory Committee.

ARTICLE X - CONFIDENTIALITY

10.1 Technical Developments and Patents shall be the sole and exclusive property of the University subject to the license rights provided under Article XI.

10.2 Monsanto shall take reasonable precautions to safeguard, in a manner comparable to that used to protect its own confidential technical information, unpublished Technical Developments and not disclose the same to others for a period of two (2) years after receipt; provided, however, that Monsanto shall not be liable for unauthorized disclosure of Technical Developments in spite of such precautions. With respect to any particular identified Technical Development for which good cause can be shown, the University may extend the two (2) year period for an additional period of two (2) years by notice in writing to Monsanto stating reasonable justification therefor and that to the University's knowledge none of the exceptions of Paragraph 10.3 is applicable. After said initial two (2) year period or extension thereof Monsanto shall be under no restrictions as to revelation of any Technical Developments. Subject to the provisions herein with respect to Patents and licenses, Monsanto shall at all times be free to use Technical Developments.

10.3 The Monsanto obligation specified in Paragraph 10.2 shall not extend to Technical Developments which:

a) become a part of the public domain or of the public knowledge through no fault of Monsanto; or

b) were in the possession of Monsanto prior to disclosure by the University, and such possession by Monsanto is documented; or

c) are received by Monsanto lawfully and properly from a third party; or

d) have been revealed in patent

applications.

10.4 Close cooperation between Monsanto personnel and University personnel in the conduct of activities required by or contributing to the purposes of this Agreement may involve the disclosure of Monsanto confidential information to such University personnel. Since, as a practical matter the University is not able to make commitments of confidentiality on behalf of its faculty nor control the confidential information disclosed to them, it shall advise all Program and Project participants that they will be required to sign in advance of receiving Monsanto confidential information personal commitments of confidentiality as Monsanto deems necessary in the circumstances.

ARTICLE XI - PATENTS AND LICENSING

11.1 Whenever the University reasonably feels a need therefor it may request Monsanto to provide in writing a preliminary indication of its current interest in commercializing Technical Developments resulting from a Project. However, Monsanto shall not be obligated to carry out commercialization.

11.2 Monsanto shall have the right and obligation to monitor progress of each Project through its representatives on the Advisory Committee and through access to University Program participants and reports, or by such other arrangements as may be mutually acceptable to Monsanto, the Program Director, and the Project Investigators as appropriate. The primary purpose of such monitoring is to detect potentially patentable inventions as early as possible. The University shall have the obligation to disclose promptly to Monsanto all potentially patentable or scientifically novel Technical Developments.

11.3 When in the judgment of Monsanto potentially patentable inventions are developed within a Project, Monsanto shall make a report of such to the University, with its views of further research that may be necessary to establish the nature and scope of these inventions, and to the extent then possible its opinion of the potential importance of such inventions to commercialization prospects, and its interests concerning the licensing by Monsanto under any Patents that may be obtained covering such inventions. The information in said report shall be retained in confidence by the University and used only for purposes of this Agreement.

11.4 When in the judgment of the University potentially patentable inventions are developed which have not yet been identified by Monsanto through the processes described in Paragraphs 11.2 and 11.3 the University shall make a report of such to Monsanto, including all available results and conclusions. Thereupon, Monsanto shall prepare and make its report to the University as specified in Paragraph 11.3.

11.5 When Monsanto has indicated its interest in a license under prospective Patent rights to an invention it shall promptly cause its patent attorneys to file and prosecute in good faith a United States Patent application on such invention. Monsanto shall also effect the filing and good faith prosecution of foreign Patent applications corresponding to the United States application in whatever countries Monsanto by written notice to the University indicates its interest in a license under prospective Patent rights.

11.6 Until such time as Monsanto notifies the University in writing that it no longer has an interest in a license, or until the expiration of the time specified in Paragraph 11.14 during which time Monsanto has not given notice of its election to take a license, Monsanto agrees to bear the cost for filing and prosecution of Patent applications under Paragraph 11.5 and the issuance and maintenance of Patents thereon. Monsanto shall not be required to prosecute any such Patent application beyond the point of final rejection by the assigned Primary Examiner in the United States Patent and Trademark Office or the equivalent stage of prosecution if a foreign application. The University, at no cost or obligation or liability to Monsanto, may take action to file or prosecute any Patent application or have issued or maintain any Patent on which Monsanto elects not to take such action. Any such election by Monsanto shall be promptly communicated to the University and in adequate time to allow the University to take such action if it so desires. Monsanto's right to a license thereunder shall not thereby be diminished.

11.7 With respect to Patent applications filed and prosecuted and Patents issued or maintained by Monsanto under Paragraphs 11.5 and 11.6, the University at its own expense may designate and retain patent counsel of its own who shall be permitted to review such Patent applications and proposed responses to Patent Office actions thereon and issuance and maintenance of Patents and to consult with Monsanto's patent attorneys before Monsanto takes action thereon. However, the control of such filings, prosecutions, issuances and maintenances shall rest with Monsanto unless it elects to relinquish such control to the University under Paragraph 11.6 by timely written notice. The University may at any time elect by notice in writing to Monsanto to assume at University's cost those activities undertaken by Monsanto under Paragraphs 11.5, 11.6 and 11.7 on behalf of the University in regard to any Patent application or Patent, and Monsanto's right to a license thereunder shall not thereby be diminished.

11.8 Title to all Patent applications and Patents issuing thereon covering Technical Developments made only by University or non-Monsanto personnel or jointly with Monsanto personnel shall be in the University. Any royalties payable with respect to the latter shall take into consideration the relative contributions of the University and Monsanto coinventors.

11.9 The parties, including the inventors, Project Investigators and Program Director, shall do all acts necessary or desirable to provide Monsanto patent attorneys with all information and records and execution of all documents necessary or desirable in the evaluation of Technical Developments, and in the filing and prosecution of Patent applications thereon, and in obtaining the issuance and maintenance of any Patents issuing from such Patent applications.

11.10 The University shall take all necessary and desirable actions, including the signing of Agreements of Program Participants (Exhibit A) by each of the persons participating in the Program, including the Program Director, all Project Investigators, and all other persons involved in the research, to assure that it acquires sufficient title to all Technical Developments, Patent applications and Patents from those of its personnel making such so as to be entitled to grant licenses to Monsanto as specified in this Agreement. The Program Director shall maintain a file of such signed Agreements of Program Participants which shall at all times be available to Monsanto representatives and upon request the Program Director shall provide Monsanto copies of specified Agreements.

11.11 In consideration of Monsanto's financial and other support of the Program and of the Patent work and cost thereof to be undertaken by Monsanto under this Article XI, the University agrees that it will make no claims against and hereby waives any claim it may have against Monsanto or its employees for injury, loss or damage resulting from acts of omission or commission by Monsanto, its employees or agents, in connection with the preparation, filing and prosecution of Patent applications and the obtaining and maintaining of Patents covering Technical Developments.

11.12 Each inventor of a potentially patentable Technical Development, no later than the time of signing a Patent application thereon, shall be requested to agree, for the considerations recited in Paragraph 11.11, to make no claims against and to waive any claims he or she may have against Monsanto or its employees for injury, loss or damage resulting from acts of omission or commission by Monsanto, its employees or agents, in connection with the preparation, filing and prosecution of Patent applications and the obtaining and maintaining of Patents covering Technical Developments. Should any inventor decline to so agree, any Patent application on such Technical Development shall be filed and prosecuted and Patents obtained and maintained by the University, at its own cost, and Monsanto's right to a license thereunder shall not thereby be diminished.

11.13 Notwithstanding any other provision of this Agreement, the University agrees to hold harmless, indemnify and defend Monsanto and its employees from all liabilities, damages, costs, expenses (including attorneys fees) and losses resulting from any claim or any lawsuit or any settlement thereof made by the University or by Monsanto with the University's consent, by the University's employees or third party having an interest through the University or its employees, and arising out of acts of omission or commission in regard to the obligations undertaken by Monsanto or its employees under Paragraphs 11.5, 11.6 and 11.7.

11.14 The University hereby agrees to grant to Monsanto licenses to make, have made, use and sell under Patents, including the right to grant sublicenses, in such countries as Monsanto may elect. Such election for any Patent shall be made within two (2) years after the filing of a Patent application in the affected country, provided, however, that Monsanto shall not be required to negotiate the terms of a license agreement until after the relevant Patent has issued.

11.15 License grants to Monsanto of rights to Patent applications and Patents issuing thereon for inventions made solely with Monsanto support shall be exclusive for the life of such Patents. For any invention made with the joint support of Monsanto and funds provided by another sponsor, or in which there is a third party inventor, such license shall, whenever legally possible, be exclusive for the life of the Patents. However, if the University is unable to grant a license which shall be exclusive for the life of the Patent, then the University shall provide Monsanto with the maximum rights permitted by law.
In connection with the transfer of Patent rights to be negotiated

under this Agreement the parties shall consider the benefits relative to licensing as distinguished from transfer of title.

11.16 The University agrees to grant and hereby grants to Monsanto an irrevocable, world-wide, paid-up, non-exclusive license, to make, have made, use and sell, including the right to grant sublicenses, on all Technical Developments which are not covered by Patents. Monsanto agrees to indemnify the University for liability arising from use of Technical Developments licensed under this Paragraph 11.16, and from use, sale or other disposition of products made by use of the said Technical Developments, by Monsanto, its affiliates, sublicensees or any party acting on behalf of same. This provision shall survive termination of this Agreement.

11.17 The University agrees to grant to Monsanto licenses on patents secured outside the Program to the extent the University has the right to so license and to the extent necessary for Monsanto to practice Technical Developments. For such patents the grant shall be on terms and conditions reasonable in the circumstances and shall include the right to grant sublicenses. Monsanto agrees to indemnify the University for liability arising from use of such patents licensed under this Paragraph 11.17 and from use, sale or other disposition of products made by use of such patents, by Monsanto, its affiliates, sublicensees or any party acting on behalf of same; this provision shall survive termination of this Agreement.

11.18 License grants to Monsanto under Paragraphs 11.14 and 11.15 shall contain at least the following terms and conditions:

a) requirement that Monsanto by its own efforts or through sublicensees during the period of exclusivity make reasonable efforts to effect the lawful introduction of Licensed Products into the marketplace as early as practicable, consistent with Monsanto's sound and reasonable business practice and judgment. The requirement for introduction of a Licensed Product into the marketplace shall be deemed met if, in the exercise of Monsanto's sound and reasonable business practice and judgment, an alternative product serving essentially the same function has been introduced into the marketplace by Monsanto and with essentially the same benefits to the consuming public.

b) the right of the University to require Monsanto to grant a non-exclusive sublicense to a responsible party on fair and reasonable terms and conditions in the event the requirement of subparagraph

a) above is not met.

c) requirement that during the period of exclusivity Monsanto submit a product development plan specifying its reasonable estimate of the schedule of key events to market entry and provide periodic reports of significant modifications to the plan and progress against the plan to the University until market entry is achieved, and requirement that the University retain in confidence the information in said plan and reports and use only for purposes of the license.

d) right of Monsanto to sublicense others provided the University is notified to whom the sublicense was granted.

e) a royalty schedule based on net selling price of Licensed Product sold by Monsanto or its sublicensees. The University and Monsanto recognize that patent protection is only one factor contributing to commercial success of a product or process and that other factors, for example other patented inventions, unpatented know-how, technical and marketing skills, financial contribution and risk, nature and extent of market, nature and extent of competition, normal trade practices, and condition of the economy also play an important part. Accordingly, rather than attempt at this time to establish royalty rates, the University and Monsanto declare their intentions to negotiate in good faith at the time of licensing, reasonable and fair royalties payable to the University by Monsanto on the commercial practice by Monsanto and its sublicensees of each Technical Development covered by a Patent licensed under this Article XI, taking into account the various factors contributing to the commercialization. If the University and Monsanto are unable to agree on royalty rates within six (6) months of the commencement of negotiation, the matter may be submitted to arbitration by either party and if so submitted by either party, shall be finally settled by arbitration conducted in accordance with the then-existing rules of conciliation and arbitration of the American Arbitration Association. Any such arbitration shall take place in St. Louis County, Missouri, before three (3) arbitrators, one of whom shall be designated by Monsanto, one by the University and the third by the two so designated. If one party fails to designate an arbitrator within thirty (30) days after the designation of an arbitrator by the other party, the arbitrator who should have been chosen by the other party shall be appointed by the American Arbitration Association as soon as possible. In the event that the said two arbitrators designated by the parties are unable to agree upon a third arbitrator within thirty (30) days after the nomination of the last of the said two arbitrators, the third arbitrator shall be appointed by the American Arbitration Association as soon as possible. None of the arbitrators need be designated from any panel published by the American Arbitration Association or any other arbitration association. The arbitrators shall apply the laws of the

State of Missouri. The decision by the arbitrators shall be binding and conclusive upon the parties, their successors and assigns and they shall comply with such decision in good faith. The University and Monsanto each shall pay its own costs and one-half of the costs of the arbitration.

f) provision that when a Licensed Product is sold but not as such and constitutes significantly less than all of the thing sold, an equitable adjustment shall be made in the net selling price of the thing sold to arrive at the net selling price for royalty calculations. When a Licensed Product is manufactured by or used in a process and the process is only a minor factor in the manufacture or use, an equitable adjustment shall be made in the net selling price.

g) provision that Monsanto payments required to be made to a third party for the right under a third-party dominating patent to make, use or sell a Licensed Product licensed hereunder shall be credited against one-half of the royalties due the University hereunder from sales of the same Licensed Product.

h) right of annual audit to confirm royalties on behalf of the University by a firm of accountants to which Monsanto has no reasonable objection.

i) indemnification of the University by Monsanto for liability arising from the manufacture, use, sale or other disposition of Licensed Products, by Monsanto or its affiliates, sublicensees or any party acting on behalf of same. This provision is to survive termination of the license agreement.

j) law of Missouri shall apply.

k) such other provisions as the parties may mutually desire, and, in the case of an exclusive license of an invention jointly supported by the government, such provisions as the government may have validly required the University to include.

1) Patent Infringement procedures:

(1) If at any time a third party shall infringe a Patent licensed to Monsanto hereunder, then Monsanto may either (i) obtain a discontinuance of such infringing operations; (ii) bring suit at Monsanto's expense against such infringer in the name of Monsanto, or in the name of the University and Monsanto if the University is a legally indispensable party; or (iii) permit the University at its option to bring such suit at its own expense. The party who brings suit shall control the prosecution and any settlements therof, and the other party shall be entitled to be represented therein by counsel of its own selection at its own expense.

(2) From any recovery from such suit or settlement thereof there shall first be paid the expenses of the party

bringing the suit, then the expenses of the other party hereto if represented by counsel, and the balance shall be divided two-thirds to the party bringing the suit and one-third to the other party, unless the parties agree otherwise.

(3) Before bringing suit Monsanto shall fully inform the University, and give careful consideration to the views of the University in making its decision whether or not to sue.

(4) If Monsanto decides to sue and University is a legally indispensble party, the University shall have the right to assign to Monsanto all of the University's rights, title and interest in the Patent or Patents concerned, in which event suit by Monsanto on such Patent or Patents shall thereafter be brought or continued solely in its name if the University is no longer an indispensable party. Patents so assigned by the University to

Monsanto shall remain subject to the same royalty and all other terms and conditions of this Agreement.

11.19 Commencing with the fourth and subsequent years in which royalties are due to the University pursuant to licenses contemplated under this agreement, Monsanto shall be entitled to a credit, not to exceed 25% of the gross royalties due for the commercialization of Licensed Products in each year, (a) of Monsanto's cumulative out-of-pocket costs (excluding the costs of Monsanto's employees) for patent activities under Paragraphs 11.5 and 11.6 and (b) 50% of all payments made prior to the date of crediting by Monsanto to the University under Article VIII hereof, which payments can be related to the cost of development of those commercialized Licensed Products.

11.20 Should Monsanto not indicate interest to take a particular license from the University, or subsequently decide not to enter into the license agreement, or terminate the license agreement, or should such agreement be justifiably terminated by the University without challenge or objection by Monsanto, then the University shall be free to license to others the subject matter so released, without further obligation to Monsanto. However, such licenses to others shall exclude Licensed Products directly competitive with or substantially equivalent to those Monsanto has licensed.

11.21 Upon the indication by Monsanto of an interest in any Technical Developments and that Monsanto desires to commence activities directed at transferring such technology to its laboratories, then the Program Director shall participate with Monsanto representatives, the Project Investigators and others as may be appropriate to work out mutually acceptable actions to be taken to effect such technology transfer, including activities contemplated under Paragraphs 6.2 and 9.6, all at no added cost to Monsanto.

ARTICLE XII - TERMINATION

12.1 This Agreement shall terminate on June 30, 1987 unless extended by mutual agreement of the parties under the provisions of Paragraph 3.2; or unless earlier terminated under the provisions of Paragraphs 4.3, 12.2 or 12.3.

12.2 In the event that either party to this Agreement defaults or breaches any of the provisions hereof, the other party reserves the right to terminate this Agreement upon ninety (90) days written notice to the defaulting party; provided that if the defaulting party, within said ninety (90) day period cures the said default or breach, this Agreement shall continue in full force and effect.

12.3 If either party shall become insolvent, or shall make any assignment for the benefit of creditors, or shall be adjudged bankrupt, or if a receiver or trustee of the property of either party is appointed, the other party on thirty (30) days written notice may terminate this Agreement.

12.4 Notwithstanding the termination of this Agreement for any reason, the provisions of Articles X, XI and XIII shall remain in effect subject to Paragraph 12.5.

12.5 If the University exercises its rights under Paragraphs 12.2 or 12.3 and validly effects the termination of this Agreement it shall be under no further obligation to grant further licenses to Monsanto and Monsanto shall promptly transfer to the University the prosecution of all pending Patent applications and the maintenance of all Patents not yet licensed to Monsanto and which Monsanto is prosecuting or maintaining hereunder.

ARTICLE XIII - INDEMNIFICATION

13.1 Monsanto agrees to hold harmless, indemnify and defend the University from all liabilities, demands, damages, expenses and losses arising out of use by Monsanto or by any third party acting on behalf of or under authorization from Monsanto, of information or materials received from University or out of any use, sale or other disposition by Monsanto or by any third party acting on behalf of or under authorization from Monsanto of products made by use of information or materials received from University.

13.2 The University warrants that it carries sufficient Worker's Compensation insurance to comply with the laws of Missouri and any other state where any of the work pursuant to this Agreement is performed with respect to the University's personnel. Except as provided under Paragraph 13.3 it is expressly understood and agreed that Monsanto shall not be responsible for or obligated in any manner to reimburse the University or to pay any compensatory, special, exemplary or consequential or other direct or indirect damages in respect of any loss, property damage, personal injuries or loss of life incurred in performance of the research work under this Agreement other than that attributable in whole or in part to Monsanto's fault or negligence, and the University shall defend, indemnify and hold Monsanto harmless (using funds other than those paid to University pursuant to the provisions of Article VIII hereof) from any and all claims, costs or liability for any such loss, damage, injuries or loss of life, other than that attributable in whole or in part to Monsanto's fault or negligence.

13.3 Monsanto agrees to defend, indemnify and hold the University harmless from any and all claims, costs or liability for any loss, damage, injury or loss of life, other than that attributable in whole or in part to the University's fault or negligence, arising as a result of any Monsanto Employee working in the laboratories of the University as provided under Paragraph 6.2.

ARTICLE XIV - TRANSFER OF INTEREST

14.1 Neither this Agreement, nor any of the rights and obligations stated herein or resulting therefrom, may be assigned, transferred or otherwise disposed of by either party without the prior written consent of the other unless such assignment, transfer or disposition is to a successor to all the business of the transferor which pertain to the subject matter of this Agreement, and provided that such successor shall agree in writing with the other party to assume all the obligations of the transferor to the other party.

14.2 Should it become necessary or desirable for the University to subcontract any of the Program research to others, such research shall be performed under a formal subcontract satisfactory to Monsanto by which the subcontractor accepts all appropriate provisions of this Agreement and other such provisions as are necessary.

ARTICLE XV - NOTICE

15.1 Any notice or report required or permitted to be given under provisions of this Agreement shall be in writing and be sent by first class mail or hand delivered:

a) If to Monsanto, to:

G. Edward Paget, M.D.
Director, Health Care Development
Monsanto Company, 02F
800 North Lindbergh Boulevard
St. Louis, Missouri 63167

with a copy to:

Mr. John E. Maurer
General Patent Counsel
Monsanto Company, E2NA
800 North Lindbergh Boulevard
St. Louis, Missouri 63167

b) If to the University, to:

David M. Kipnis, M.D.
Chairman, Department of Medicine
Washington University School of Medicine
660 South Euclid Avenue
St. Louis, Missouri 63110

with a copy to:

Mr. Edward L. MacCordy
Associate Vice Chancellor for Research
Washington University
Lindell & Skinker Boulevards
St. Louis, Missouri 63130

15.2 Either party may change the address or the person(s) designated to receive notice by notifying the other in writing of the change.

ARTICLE XVI - GENERAL PROVISIONS

16.1 Except as provided in Paragraph 9.5, neither party shall use the name of the other party, its affiliated organizations or its personnel in advertising or promotional materials or news or press releases pertaining to the subject matter of this Agreement without prior written consent of such other party.

16.2 This Agreement shall be construed under the laws of the State of Missouri.

16.3 No waiver of any default, condition, provisions or breach of this Agreement shall be deemed to imply or constitute a waiver of any other like default, condition, provision or breach of this Agreement.

16.4 The Article headings used in this Agreement are for convenience only and form no part of the Agreement.

16.5 This writing constitutes the entire Agreement between the parties hereto relating to the subject matter of this Agreement and there are no understandings, representations or warranties of any kind except as expressly provided herein. Neither this Agreement, nor any term or provision thereof, may be discharged, waived, released, abandoned, changed or modified except by an instrument in writing signed by a duly authorized representative of each of the parties to this Agreement. If either party desires a modification or change of any kind in this Agreement, the parties shall, upon reasonable notice of the proposed modification or change by the party desiring the change, confer in good faith to determine the desirability of such modification or change.

16.6 The parties agree that it is the intention of neither party to violate any valid federal, state and local laws and regulations; that if any sentence, paragraph, clause, or combination of the same in this Agreement is in violation of any applicable law or regulation, or is unenforceable or void for any reason whatsoever, such sentence, paragraph, clause or combinations of the same shall be inoperative and the remainder of the Agreement shall remain binding upon the parties.

IN WITNESS WHEREOF, the parties have caused this Agreement to be executed in duplicate by their duly qualified officers.

THIS CONTRACT CONTAINS A BINDING ARBITRATION PROVISION WHICH MAY BE ENFORCED BY THE PARTIES.

WASHINGTON UNIVERSITY

By Wm Danforth
William H. Danforth
Chancellor

Date 6/1/82

MONSANTO COMPANY

By
John W. Hanley
Chairman of the Board

Date 6/3/82

EXHIBIT A

AGREEMENT OF PROGRAM PARTICIPANTS

The purpose of the following agreement is to describe the responsibilities of and to enlist the support and cooperation of research participants and to insure compliance with relevant University policies.

Therefore, as a participant in a research project under the Biomedical Research Program sponsored and funded by Monsanto Company, I agree to abide by the following terms and conditions:

1. PATENTABLE INVENTIONS:

(a) Participants will promptly disclose to the University's Program Director any potentially patentable invention or novel scientific development they produce in any research Project funded by Monsanto. Such disclosure will occur prior to disclosure to any other non-Program participant.

(b) Participants will, upon request, assign rights to patentable inventions to the University so that it may grant required licenses to the sponsor.

(c) Participant inventors will cooperate with Monsanto and

University patent attorneys in the filing and prosecution of patent applications. Due to the major expense and specialized professional assistance required to pursue patent rights in a research program of this magnitude, Monsanto has assumed this responsibility. The University will monitor these efforts and at its option may assume such responsibility on a case by case basis.

(d) In consideration of Monsanto's willingness to file and prosecute patent applications at its own expense, participant inventors will be requested to waive any claim of liability by Monsanto in these efforts. Otherwise, the University must assume this responsibility and its expense.

(e) Any royalties from licensed inventions received by the University will be distributed as follows: 40% to the research laboratory(ies) in which the invention was made, 40% to the cognizant department(s), and 20% to the School of Medicine.

2. PRODUCTS OF RESEARCH:

(a) New materials, processes, devices, scientific information, and any other research products isolated or developed in a project, whether patentable or not, will be made available to Monsanto for its evaluation and general use.

(b) Such research products may be made available to other research scientists at non-profit institutions according to normal academic practice. However, recipient scientists should agree not to further distribute such research products and not to use them for the benefit of another commercial firm. Distribution of potentially patentable research products should not be made until Monsanto has evaluated patentability and, if appropriate, filed a patent application.

3. <u>PUBLICATIONS</u>:

(a) Scientific advances made under this research program will be freely reported in the scientific literature.

(b) Two (2) copies of each proposed publication, including abstracts, in the best form then available will be provided to the Program Director at least one (1) month before being submitted for publication.

(c) Based on a review by Monsanto patent attorneys of the proposed article, a brief delay in its submission for publication may be necessary to allow the filing of adequate patent applications. Such brief delay may occasionally be necessary to avoid the loss of patent rights.

(d) Two (2) copies of the final abstract and article as submitted to the publisher shall be simultaneously provided to the Program Director.

(e) Each publication will acknowledge Monsanto Company support of the research being reported.

(f) Prior to the evaluation of research results for potentially patentable inventions, participants will use caution in public or other outside presentations and discussions not to prematurely disclose critical technical information which could result in the loss patent rights.

4. COOPERATION WITH MONSANTO:

(a) It is intended that there be mutually productive and continual interchange between the University and Monsanto scientists. For this purpose a Monsanto Project Scientist will be appointed as the primary company contact with each Project Investigator. Each Project Investigator will be available for consultation with the Monsanto Project Scientist on matters concerning the project.

(b) These University and Monsanto scientists will, as necessary, identify Monsanto special facilities and capabilities which may be used by the Project Investigator to enhance the progress of his/her project.

(c) Project Investigators will, upon request by Monsanto, provide reasonable opportunities for individual Monsanto scientists and technicians to spend time in the research laboratories to learn newly developed techniques, to participate in the research if this is mutually desirable, and to assist in the transfer of newly developed technology to Monsanto.

(d) The cooperative nature of this research program is expected to necessitate the exposure of University participants to Monsanto confidential technical information. For participants who may be so exposed Monsanto will require in advance the signing of a personal agreement indicating the partcipants willingness not to disclose such Monsanto confidential information to others.

5. <u>AVOIDING CONFLICT SITUATIONS</u>:

(a) Participants in research projects under this program must consider all other activities in which they are engaged, or have a personal interest, or in which they may become involved during the term of their project so that they reasonably avoid conflicting obligations. Of special concern are obligations to other companies in the same scientific areas or closely related to their research work supported by Monsanto. This project should not overlap the research they are performing or plan to perform under the sponsorship of any other organization, including government agencies and foundations, unless the situation is known to and approved by the Program Director.

(b) Any potential conflict of obligations or interests faced by a participant involving a proposed or approved project under this program must be promptly disclosed to the Program Director.

(c) The Program Director may request disclosure by project personnel of their past, current or anticipated relationships with other organizations in order to assure the absence of possible conflicts.

6. <u>PROGRESS REPORTS</u>:

In order for Monsanto to be fully informed about research results and to be able to identify potentially patentable inventions as early as possible, occasional brief summary reports of important findings and results will be required of

Project Investigators, as will more detailed annual progress reports which include summaries and conclusions.

The above terms and conditions are understood and agreed to:

P.I. Typed Name	____________	Other Project Personnel:	
Signature	____________	Sig.	____________
Date	____________	Sig.	____________
Phone No.	____________	Sig.	____________

Appendix D
Washington University-Anheuser-Busch Companies, Inc., Micromixing Research Agreement of 1983

WHEREAS, Anheuser-Busch Companies, Inc., a Delaware corporation with principal offices at One Busch Place, St. Louis, Missouri 63118 (hereinafter "ABC"), through one or more of its subsidiary companies, is engaged in research, production and sale of consumer food products; and

WHEREAS, the production of certain of ABC's consumer food products involve fermentation technology; and

WHEREAS, ABC is desirous to research, have researched and participate in research in the area of biotechnology and biochemical engineering as particularly pertaining to fermentation and to apply the results of such research in its business; and

WHEREAS, Washington University, a corporation organized under the laws of the State of Missouri, with principal offices at Lindell and Skinker Boulevards, St. Louis, Missouri 63130 (hereinafter "UNIVERSITY"), through its Center for Biotechnology (hereinafter "CENTER") and under the particular leadership of Eric H. Dunlop, Ph.D. (hereinafter "DUNLOP") is prepared and willing to undertake and have undertaken biotechnological and biochemical engineering research and development in the specific area of **"micromixing"** (hereinafter more specifically defined); and

WHEREAS, ABC, UNIVERSITY and DUNLOP are desirous to engage in a high degree of cooperation in connection with the research to be conducted,

NOW THEREFORE ABC, UNIVERSITY and DUNLOP, in consideration of the mutual promises hereinafter made, agree as follows:

1. "MICROMIXING" - DEFINITION

The parties understand the term "Micromixing" (hereinafter "MMX") to encompass advanced biochemical engineering technology in connection with the mixing of fermentation substances on a micro-scale, comparable in size to a yeast cell, with a view toward more efficient and effective ways to provide nutrients to and remove metabolic waste from yeast.

2. RESEARCH SUBJECTS

Subject to Section 9 (NO CONFLICTS) herein, UNIVERSITY agrees to have conducted, through its CENTER, studies and research in the area of MMX, as this term has hereinbefore been defined, so as to improve and facilitate the mixing of fermentation substances in production and operational situations. The parties agree that - given additional research and study efforts - much improvement and economic benefits may be derived in the industrial application of MMX. The research and development effort shall be specifically directed to achieve such improvements and shall be conducted by the CENTER, under the specific leadership of DUNLOP as hereinafter more particularly set forth.

ABC acknowledges a high regard for the CENTER and particularly for Dr. Eric H. DUNLOP. ABC has been induced to enter into this Agreement because of the particular expertise and academic credentials of DUNLOP. Accordingly, the parties agree that all research and development efforts shall be conducted under the direct supervision and with considerable involvement of DUNLOP. The parties recognize that certain aspects of the research can be conducted and performed by persons other than DUNLOP, but under DUNLOP's direction and supervision. UNIVERSITY and DUNLOP agree to involve only appropriately qualified graduate students, research scientists and technicians. UNIVERSITY further agrees to allow, make available, and require DUNLOP and DUNLOP agrees to devote not less than 15% of his time, on a yearly basis, on the MMX research project.

3. FINANCIAL TERMS

Subject to the provisions of paragraphs 5, 6 and 12 herein, ABC agrees to pay to UNIVERSITY, for the research and development to be conducted, a grant in the aggregate sum of $400,000 payable in four yearly instalments of $100,000 each, beginning with the date this Grant Agreement is executed and upon each of the succeeding three anniversary dates thereof.

It is agreed that the grant money is intended for the support of graduate students, research scientists and technicians, for the purchase of supplies and equipment, to defray expenses for the attendance at scientific meetings, for indirect costs at the approved, federally negotiated rate and for other expenditures reasonably incurred by the CENTER. It is further agreed that the CENTER shall control the funds subject, however, to DUNLOP's specific approval over the usage and disbursement thereof. The parties also agree to permit periodic review regarding allocation of the grant funds by the Joint Advisory Committee pursuant to paragraph 5. Title to all equipment shall vest in the UNIVERSITY upon purchase.

4. TERM

The term of this Agreement, except as provided in paragraphs 10 and 15, shall be for four years beginning with the date of its execution and terminating upon the fourth anniversary date thereof. The parties understand that it is impossible to predict with certainty the precise length of time or amount of effort, resources, and funding which may be necessary to develop the technology of the MMX research effort to a level of meaningful practical application. While the parties are optimistic that some practical application will be achieved during the term of this Agreement, it is their stated desire to mutually assess the success of the research program at the conclusion of the term of this Agreement and evaluate, at that time, the advisability of renewing or extending this Agreement under such terms and conditions as may be desirable or appropriate.

5. JOINT ADVISORY COMMITTEE

The parties recognize and agree that the research and development efforts be conducted with a high degree of coordination, mutual trust, and interfacing between and amongst their respective technical representatives. To that end, the parties agree to establish a Joint Advisory Committee (hereinafter "COMMITTEE") to be comprised of three representatives of UNIVERSITY, to include DUNLOP as chairman, and three representatives of ABC. The parties agree the purposes of the COMMITTEE to be the mutual assessment of research being conducted and to be conducted, discussions of ad hoc problems and resolutions thereof, scientific interfacing as necessary or indicated and all other matters of mutual concern in connection with the research effort envisioned by this Agreement.

The COMMITTEE shall meet regularly, but not less than once every three months at a location or locations to be agreed upon. In the event of "Significant Changes" as described in paragraph 12 herein, the COMMITTEE shall meet specially and promptly to address problems associated therewith.

The COMMITTEE shall, at its quarterly meetings, also review how the grant fund is being allocated by CENTER and DUNLOP. If warranted by exceptional progress or promise, the COMMITTEE may recommend to ABC that the funding be expanded beyond the terms of Paragraph 3 (FINANCIAL TERMS) herein. The COMMITTEE may also recommend to diminish or redirect to another project altogether, any funding not expended. In all cases, however, ABC will make the final decision, taking into consideration the recommendations of the COMMITTEE.

6. PROGRESS REPORTS

DUNLOP shall provide ABC written progress reports from time to time so as to permit ABC to become knowledgeable with and evaluate the research as it progresses. Such progress reports shall be submitted to ABC at such frequencies and at such times so as to coincide with and forming the basis of quarterly meetings of the COMMITTEE. In addition, DUNLOP shall provide ABC with a comprehensive annual Progress Report in June of each year during the term of this Agreement. Such annual Progress Reports shall include a relative comparison between grant monies given by ABC and grant monies spent or allocated by CENTER and DUNLOP.

7. EQUIPMENT

UNIVERSITY agrees to make available, through its CENTER, all necessary equipment and materials for the research endeavor. The parties recognize, however, that certain equipment and materials, suitable for research and testing, may be located upon the premises of and owned by ABC. To the extent that the usage of such equipment and materials may be deemed necessary or desirable, ABC will make same available to DUNLOP and CENTER's researchers, provided that such usage does not unreasonably interfere with ABC's production activities.

8. CONFIDENTIALITY

The parties hereby acknowledge that materials and information (except information derived from ABC) disclosed in requests for project funding whether subsequently approved or not, in project progress reports whether written or oral, in annual reports, in discussions in COMMITTEE meetings and between project participants and representatives of the ABC, in invention disclosures to ABC, and in abstracts and articles proposed for publication until published, constitute valuable UNIVERSITY property. Accordingly, ABC agrees that it shall take reasonable precautions to safeguard such information and materials in a manner comparable to that used to protect its own confidential technical information and materials. However, ABC shall not be liable for unauthorized revelation which occurs in spite of such precautions.

ABC's obligation specified above shall not extend to information or materials which:

(a) become a part of the public domain or of the public knowledge through no fault of ABC; or

(b) were in the possession of the ABC prior to disclosure by the UNIVERSITY, and such possession by the ABC is documented; or

(c) are received by ABC lawfully and properly from a third party; or

(d) must be revealed in patent applications.

Subject to the above exceptions ABC agrees to restrict dissemination of information and materials to only those ABC employees and others whose knowledge thereof is reasonably necessary or useful to ABC's pursuit of the purposes of this Agreement.

Close cooperation between ABC personnel and UNIVERSITY personnel in the conduct of activities required by or contributing to the purposes of this Agreement may involve the disclosure of ABC confidential information to such UNIVERSITY personnel. Since, as a practical matter the UNIVERSITY is not able to make commitments of confidentiality on behalf of its faculty nor control the confidential information disclosed to them, it shall advise all project participants that they will be required to sign in advance of receiving ABC confidential information personal commitments of confidentiality as ABC deems necessary in the circumstances.

The parties recognize that the collaborative research, as envisioned by this Agreement, may result in researchers and scientists of UNIVERSITY having access to confidential or secret information of ABC and ABC's research representatives having access to confidential information of UNIVERSITY. To properly safeguard such information, the parties agree to execute personal confidential non-disclosure agreements if deemed necessary and appropriate by the requesting party.

9. NO CONFLICT

The parties understand that general research in the field of MMX may be beneficial to ABC and others. In recognition of that fact, CENTER and/or DUNLOP may accept - concurrent research opportunities, consultation, grants or contracts generally relating to MMX, unless, however, such research, consultation, grants or contracts occur in the specific areas of interest to ABC, namely beer brewing, wine making, baking or the production of baker's yeast and all other yeast products, including recombinant DNA yeast and other genetically modified yeasts and their products. CENTER and DUNLOP specifically agree not to accept consultation or grant assignments from any other party in connection with research in the area of MMX which is in conflict with the hereinbefore enumerated specific areas of interest to ABC during the term of this Agreement without the prior written consent of ABC.

UNIVERSITY and DUNLOP specifically state that the research to be conducted under this grant in the area of MMX, as envisioned by this Agreement, does not conflict with any existing research being conducted by CENTER or DUNLOP at the time of the execution of this Agreement.

In the event governmental, corporate or individual sponsors for concurrent or additional research in the area of MMX should come forward, then, in such event, the COMMITTEE shall meet to discuss and review any such potential sponsorship and make a recommendation to ABC as to whether any conflict exists with ABC's specific areas of interest. The final decision as to whether a conflict exists with ABC's specific areas of interest shall rest with ABC.

10. PATENTS AND LICENSES

In the event that inventions and discoveries are made which warrant the pursuit of Letters Patent, such inventions and discoveries shall be assigned by the inventors of same to UNIVERSITY, and UNIVERSITY shall undertake to obtain patentability searches and, if justified from the searches, to arrange for the preparation, filing and prosecution of patent applications.

ABC shall reimburse to UNIVERSITY reasonable costs of patent searches, and other out-of-pocket costs and expenses in the drafting and prosecution of patent applications and in the issuance of any Letters Patent granted on said applications; provided, however, that ABC reserves the right to review the entire prosecution record of any patent applications filed and ABC reserves the right of approval of any instructions to attorneys or to the Patent and Trademark Office which are likely to incur a reimbursable cost or expense.

In consideration of ABC's making the above-described costs and expenses, as well as the grant payments outlined in Paragraph 3 of this Agreement, ABC shall have an exclusive royalty-free license in perpetuity with the right to sublicense to make, use and sell any process, product, composition of matter, article of manufacture, or improvement of same which is forthcoming in this research and development program, whether patentable or not.

However, notwitstanding the above, ABC shall have no longer than a period of 10 years from the date of receiving the initial disclosure of any such process, product, composition of matter, article of manufacture or improvement in which to evaluate the commercial feasibility of same and to either sublicense it or incorporate it into ABC's own commercial operations. At the end of such 10-year period, if ABC has failed to take any action with respect to sublicensing or incorporating such inventions or discoveries into its own commerical operations, ABC's license shall become nonexclusive.

In the event ABC sublicenses a third party to make use of or sell any invention or discover herein and collects a royalty from the third party licensee, UNIVERSITY shall receive 50 percent of said royalty.

Neither UNIVERSITY nor DUNLOP will withhold information regarding this program, or in any way serve to inhibit or interfere with ABC's right of access to all information, inventions and discoveries growing out of the research.

11. PUBLICATIONS:

Any scientific advances made during the course of the research may be freely reported by DUNLOP and/or UNIVERSITY. Before any such report is made, however, an abstract thereof shall first be submitted to ABC for review. ABC shall have 30 days within which to either approve the report or request a delay in reporting. ABC shall have the right to delay publication for a reasonable period of time in order to secure intellectual property rights. Publication of information shall be permitted at any time, however, following the filing of a patent application claiming the discovery sought to be published.

12. SIGNIFICANT CHANGES

Inasmuch as ABC, as indicated in paragraph 2 hereof, is relying upon the particular expertise of DUNLOP to oversee and foster the advancement of the research in the stated and defined area of MMX, the parties acknowledge that DUNLOP's absence from or other non-availability to the research project is critical. In the event, therefore, of any non-availability of DUNLOP for a period of time in excess of two (2) months, the COMMITTEE shall meet as soon as practical to discuss the impact of DUNLOP's non-availability upon the research project and make recommendations regarding and the disposition of the research project as well as the continued funding thereof. The final decision shall rest, however, with ABC.

13. INDEMNIFICATION

UNIVERSITY shall hold ABC harmless and indemnify ABC in connection with any losses, damages, or liabilities arising directly or indirectly out of the negligence of UNIVERSITY, its employees or agents when engaged in research activities contemplated by this Agreement.

ABC shall hold harmless and indemnify UNIVERSITY from any losses, damages or liabilities arising directly or indirectly out of the negligence of ABC, its employees or agents when engaged in research activities contemplated by this Agreement.

ABC agrees to indemnify, defend and hold harmless UNIVERSITY from all claims, damages and losses arising out of the use or sale of any product developed from or based upon information or materials received from UNIVERSITY under this Agreement.

14. FORCE MAJEURE

Notwithstanding any other provision in this Agreement, the parties agree that to the extent either party is precluded from performing its obligations under this

Agreement by reason of circumstances beyond their reasonable control, the parties shall be excused, to such extent, from performing their obligations. In the event of such circumstances, the parties will make every effort within their control to resume fulfillment of their respective obligations as expeditiously as possible.

15. EARLY TERMINATION AND RENEWAL

Notwithstanding the provisions of paragraph 12 (SIGNIFICANT CHANGES) herein, in the event UNIVERSITY and/or DUNLOP shall, during the term of this Agreement, discontinue to conduct or have conducted the research forming the basis of this Agreement and such discontinuance is not excused under the provisions of paragraph 14 (FORCE MAJEURE) herein, then, in such event, and upon written notice to UNIVERSITY and DUNLOP, ABC shall have the option to terminate this Agreement. In such event, ABC shall be relieved from its obligation to pay to UNIVERSITY any grant money instalments not then due and UNIVERSITY and DUNLOP agree to return any unspent funds to ABC. In keeping with the cooperative spirit with which the parties have entered into this Agreement it is, however, the stated intention of the parties that before ABC exercises its option to terminate this Agreement under this paragraph, the parties will mutually discuss, through the COMMITTEE, any apparent discontinuance on the part of UNIVERSITY and/or DUNLOP with the research project.

In the event that ABC applies any information, inventions, or discoveries from this research to commercial practice within ten (10) years from its receiving the intial disclosure of same and this application reduces ABC's costs or increases ABC's net profits by an amount greater than $400,000, ABC agrees to provide an additional $400,000 grant to the CENTER.

In addition, any party may terminate this Agreement during the initial four-year term by giving, in writing, notification twelve months in advance.

16. GOVERNING LAWS

This Agreement shall be governed and construed in accordance with the laws of the State of Missouri.

17. NOTICES

Any written notices required herein shall be sent registered mail, addressed to the parties as follows:

If to UNIVERSITY:
Washington University
Lindell and Skinker Blvds.
St. Louis, MO 63130
Attn: Director, Research Contracts and Licensing Administration

If to DUNLOP:
Dr. Eric H. Dunlop
Center for Biotechnology
Washington University
Lindell and Skinker Blvds.
St. Louis, MO 63130

If to ABC:
Anheuser-Busch Companies, Inc.
One Busch Place
St. Louis, MO 63118
Attn: Director, Corporate Research and Development

18. ASSIGNMENTS

Neither party to this Agreement shall assign their respective obligations or interests hereunder to any other party. With the exception, as stated in this Agreement, of allowing qualified graduate students, research scientists and technicians to participate in the research projects, UNIVERSITY or DUNLOP shall not subcontract any of their obligations under this Agreement.

19. ENTIRE AGREEMENT

This Agreement shall constitute the entire agreement between the parties. There are no promises, terms, conditions, or obligations except those contained herein and this Agreement shall supersede any and all previous communications, representations, agreements, either verbal or written, between the parties.

20. AMENDMENTS

This Agreement may only be modified or amended by a written instrument signed by the authorized representatives of a party hereof.

21. HEADINGS

The paragraph headings appearing in this Agreement have been inserted for the purpose of convenience and ready reference. They do not purport, and shall not be deemed, to define, limit or extend the scope or intent of the paragraphs to which they appertain.

IN WITNESS WHEREOF, UNIVERSITY, DUNLOP, and ABC executed this Agreement this 15th day of August, 1983.

Attest: Harriet R. Fortier

WASHINGTON UNIVERSITY

By: [signature]

Title: Associate Vice Chancellor for Research

[signature]

ERIC H. DUNLOP, Ph.D.

Attest: [signature]

ANHEUSER-BUSCH COMPANIES, INC.

By: [signature]

Title: Vice President & Group Executive

Selected Bibliography

BOOKS AND MONOGRAPHS

Allen, Thomas J. *Managing the Flow of Technology: Technology Transfer and the Dissemination of Technological Information within the R&D Organization*. Cambridge, Mass.: MIT Press, 1977.

Association of American Universities. *The Nation's Deteriorating University Research Facilities: A Survey of Recent Expenditures and Projected Needs in Fifteen Universities—Executive Summary*. Washington, D.C.: Association of American Universities, July 1981.

Baer, Walter S. "University-Industry Interactions in Civil Sector R&D." Rand Corporation, Santa Monica, Calif., P–5724, October 1976.

———. "University Relationships with Other R&D Performers." Rand Corporation, Santa Monica, California, P–5780, January 1977.

———. "The Changing Relationships: Universities and Other R & D Performers." In *The State of Academic Science: Background Papers*. Ed. B. L. R. Smith and J. J. Karlesky. New Rochelle, N.Y.: Change Magazine Press, 1978.

———. *Report on Strengthening University-Industry Interactions*. Rand Corporation for National Science Foundation, July 1979, revised January 1980.

Barfield, Claude E. *Science Policy from Ford to Reagan: Change and Continuity*. Washington, D.C.: American Enterprise Institute for Public Policy Research, 1982.

Baxter, James, III. *Scientists Against Time*. Boston: Little, Brown and Co., 1946.

Bogert, George G. and George T. *The Law of Trusts and Trustees*. St. Paul, Minn.: West Publishing Co., 1971.

Bok, Derek C. *Beyond the Ivory Tower: Social Responsibility of the Modern University*. Cambridge, Mass: Harvard University Press, 1982.

Botkin, James, Dan Dimancescu, and Ray Stata. *Global Stakes: The Future of High Technology in America*. Cambridge, Mass.: Ballingen Publishing Co., 1982.

Bozeman, Barry, Michael Crow, Albert Link, eds. *Strategic Management of Industrial R&D*. Lexington, Mass.: Lexington Books, 1984.

Branscomb, Lewis M. "Opportunities for Cooperation Between Government, Industry, and the University." *Annals of the New York Academy of Sciences* 334 (14 December 1979).

Broad, William J., and Nicholas Wade. *Betrayers of the Truth: Fraud and Deceit in the Halls of Science*. New York: Simon and Schuster, Inc., 1982.

Brodsky, N., H. G. Kaufman, and J. D. Tooker. *University Industry Cooperation:*

A Preliminary Analysis of Existing Mechanisms and their Relationship to the Innovation Process. New York: NYU Center for Science and Technology Policy, July 1979. (Under National Bureau of Standards Order No. NB79NAA/A8898.)

Callman, Rudolf. *Law of Unfair Competition, Trademarks and Monopolies*. Wilmette, Ill.: Callaghan & Co., 1985, 4th ed.

Cary, William L., and Craig B. Bright. *The Developing Law of Endowment Funds: The Law and the Lore Revisited*. New York: The Ford Foundation, 1974.

Chisum, Donald S. *Patents: A Treatise on the Law of Patentability, Validity and Infringement*. New York: Matthew Bender & Co., 1983.

Clowse, Barbara B. *Brainpower for the Cold War: The Sputnik Crisis and National Defense Education Act of 1958*. Westport, Conn.: Greenwood Press, 1981.

Cox, Donald W. *America's New Policy Makers: The Scientist's Rise to Power*. Philadelphia: Chilton Books, 1964.

Culliton, Barbara. "Reactors." In *Partners in the Research Enterprise: University-Corporate Relations in Science and Technology*. Ed. Thomas W. Langfitt, Sheldon Hackney, Alfred P. Fishman, and Albert V. Glowasky. Philadelphia: University of Pennsylvania Press, 1983.

Danforth, William H., et al. *Report of the Association of American Universities Committee on the Integrity of Research*. Washington, D.C.: Association of American Universities, 1983.

Dickson, David. *The New Politics of Science*. New York: Pantheon Books, 1984.

Dupree, A. Hunter. *Science in the Federal Government: A History of Policies and Activities to 1940*. Cambridge, Mass.: Harvard University Press, 1957.

Eddy, Edward D., Jr. *Colleges for Our Land and Time: The Land-Grant Idea in American Education*. New York: Harper & Brothers, 1957.

Encyclopaedia of the Social Sciences. S.v. "Universities and Colleges" by Stephan d'Irsay.

Fred, E. B. *The Role of the Wisconsin Alumni Research Foundation in the Support of Research at the University of Wisconsin*. Madison, Wis.: Wisconsin Alumni Research Foundation, March 1973.

Garahan, James C., James P. Fuller, and Frederick R. Chilton, Jr. "Research and Development Limited Partnerships." In *Research and Development Limited Partnerships*. New York: Practicing Law Institute, 1983.

Giamatti, A. Bartlett. *The University and the Public Interest*. New York: Atheneum, 1981.

———. "Free Market and Free Inquiry: The University, Industry, and Cooperative Research." In *Partners in the Research Enterprise: University-Corporate Relations in Science and Technology*. Ed. Thomas W. Langfitt, Sheldon Hackney, Alfred P. Fishman, and Albert V. Glowasky. Philadelphia: University of Pennsylvania Press, 1983.

Givens, R. A., ed. *Legal Strategies for Industrial Innovation*. Colorado Springs, Colo.: McGraw-Hill, 1982.

Greenberg, Daniel S. *The Politics of Pure Science*. New York: New American Library, 1967.

Grounds, Preston W. *University–Industry Interaction: Guide to Developing Fundamental Research Agreements*. Allentown, Pa.: Council for Chemical Research, 1983.

Guze, Samuel B. "The Monsanto–Washington University Biomedical Research Agreement." In *Partners in the Research Enterprise: University-Corporate Relations in Science and Technology*. Ed. Thomas W. Langfitt, Sheldon Hackney, Alfred P. Fishman, and Albert V. Glowasky. Philadelphia: University of Pennsylvania Press, 1983.

Hofstadter, Richard, and C. Dewitt Hardy. *The Development and Scope of Higher Education in the United States*. New York: Columbia University Press, 1952.

Hofstadter, Richard, and Walter Metzgar. *The Development of Academic Freedom in the United States*. New York: Columbia University Press, 1955.

Horowitz, Irving L., and James E. Katz. *Social Science and Public Policy in the United States*. New York: Praeger Publishers, 1975.

Hutchins, Robert M. *The Higher Learning in America*. New Haven, Conn.: Yale University Press, 1936 (2d printing 1962).

Industrial Research Institute, Inc. *Industrial Innovation: The Impact of Federal Policies on Industry/University Relations. A Position Statement by the Industrial Research Institute*. New York: Industrial Research Institute, 26 September 1980.

International Encyclopedia of the Social Sciences. S.v. "Research and Development: Financing Social Science Research," by Henry W. Kiecker.

Johnson, Elmima C., and Louis G. Tornatzky. "Academia and Industrial Innovation." In *New Directions for Experiential Learning: Business and Higher Education—Toward New Alliances*, pp. 47–63. Ed. by Gerard G. Gold. San Francisco: Jossey-Bass, 1981.

Johnson, P. S. *Cooperation Research in Industry*. New York: John Wiley & Sons, 1973.

Joughin, George L., ed. *Academic Freedom and Tenure*. Madison: University of Wisconsin, 1969.

Kaplin, William A. *The Law of Higher Education*. San Francisco: Jossey-Bass, 1978.

Katz, James E. *Presidential Politics and Science Policy*. New York: Praeger Publishers, 1978.

Keller, George. *Academic Strategy: The Management Revolution in American Higher Education*. Baltimore: Johns Hopkins University Press, 1983.

Kerr, Clark. *The Uses of the University*. 3d ed. Cambridge, Mass. Harvard University Press, 1982.

Kevles, Daniel J. *The Physicists: The History of the Scientific Community in Modern America*. New York: Vintage Books, 1979.

Kidd, Charles V. *American Universities and Federal Research*. Cambridge, Mass: Harvard University Press, 1959.

Killian, J. R., Jr. "University Research." In *The Corporation and the Campus*. Ed. Robert H. Connery. New York: Praeger Publishers, 1970.

Knowles, Asa S. *Handbook of College and University Administration: Academic*. New York: McGraw-Hill, 1970.

Lambright, W. Henry. *Governing Science and Technology*. New York: Oxford University Press, 1976.

Langfitt, Thomas W. "Epilogue." In *Partners in the Research Enterprise: University-Corporate Relations in Science and Technology*. Ed. Thomas W. Langfitt,

Sheldon Hackney, Alfred P. Fishman, and Albert V. Glowasky. Philadelphia: University of Pennsylvania Press, 1983.

Langfitt, Thomas W., et al. eds. *Partners in the Research Enterprise: University-Corporate Relations in Science and Technology*. Philadelphia: University of Pennsylvania Press, 1983.

Linnell, Robert H. "Ethical and Economic Issues Facing Academe in a Rapidly Changing Society." In *Dollars and Scholars: An Inquiry into the Impact of Faculty Income Upon the Function and Future of the Academy*. Ed. R. H. Linnell. Los Angeles: University of Southern California Press, 1982.

Mansfield, Edwin. *The Economics of Technological Change*. New York: W. W. Norton & Company, Inc., 1968.

———. *Industrial Research and Technological Innovation, An Econometric Analysis*. New York: W. W. Norton & Company, Inc., 1968.

———. *Technological Change, An Introduction to a Vital Area of Modern Economics*. New York: W. W. Norton & Company, Inc., 1971.

Mansfield, Edwin, et al. *Research and Innovation in the Modern Corporation*. New York: W. W. Norton & Company Inc., 1971.

Mansfield, Edwin, et al. *The Production and Application of New Industrial Technology*. New York: W. W. Norton & Company, Inc., 1977.

Mansfield, Edwin, et al. *Technology Transfer, Productivity, and Economic Policy*. New York: W. W. Norton & Company, Inc., 1982.

Matthews, Jana B., and Rolf Norgaard. *Managing the Partnership Between Higher Education and Industry*. Boulder, Colo.: National Center for Higher Education Management Systems, 1984.

Merton, Robert K. *The Sociology of Science: Theoretical and Empirical Investigations*. Chicago: University of Chicago Press, 1973.

Miller, Jon D. *The American People and Science Policy: The Role of Public Attitudes in the Policy Process*. New York: Pergamon Press, 1983.

Nash, Paul. *Authority and Freedom in Education: An Introduction to the Philosophy of Education*. New York: John Wiley & Sons, 1966.

National Commission on Research. *Accountability: Restoring the Quality of the Partnership*. Washington, D.C.: National Commission on Research, March 1980.

———. *Funding Mechanisms: Balancing Objectives and Resources in University Research*. Washington, D.C.: National Commission on Research, May 1980.

———. *Industry and the Universities: Developing Cooperative Research Relationships in the National Interest*. Pasadena, Calif.: National Commission on Research, August 1980.

Nelkin, Dorothy. *Science as Intellectual Property: Who Controls Research?* New York: Macmillan Publishing Co., 1984.

Nelson, Richard R., ed. *Government and Technical Progress, A Cross-Industry Analysis*. New York: Pergamon Press, 1982.

Noble, David F. *America by Design: Science, Technology, and the Rise of Corporate Capitalism*. New York: Alfred A. Knopf, 1977.

———. *Forces of Production: A Social History of Industrial Automation*. New York: Alfred A. Knopf, 1984.

Petillon, Lee R., and Robert Joe Hull. *R&D Partnerships: Structuring the Transaction*. New York: Clark Boardman Co., Ltd., 1984.

Price, Don K. *Government and Science: Their Dynamic Relation in American Democracy*. New York: New York University Press, 1954.

———. "Federal Money and University Research." In *Science Policy and the University*. Ed. Harold Orlans. Washington, D.C.: Brookings Institution, 1968.

Rae, John. "The Application of Science to Industry." In *The Organization of Knowledge in Modern America, 1860–1920*, pp. 249–68. Ed. Alexandra Oleson and John Voss. Baltimore: Johns Hopkins University Press, 1979.

Reich, Robert B. *The Next American Frontier: A Provocative Program for Economic Renewal*. New York: Penguin Books, 1983.

Rosenberg, Peter D. *Patent Law Fundamentals*. New York: Clark Boardman Co., Ltd., 1975.

Rosenzweig, Robert M. "Faculty and Standards of Ethical Conduct." In *The Ethics of Teaching and Scientific Research*. Ed. Sidney Hook, Paul Kurtz, and Miro Todorovich. New York: Prometheus Books, 1977.

———. *The Research Universities and Their Patrons*. Berkeley: University of California Press, 1982.

Ross, Earle D. *Democracy's Colleqe: The Land-Grant Movement in the Formative Style*. Ames, Iowa: Iowa State College Press, 1942.

Rowen, James. "Politics of University Research." In *Academic Super Markets*. Ed. Philip G. Altbach. San Francisco: Jossey-Bass, 1971.

Saliwanchik, Roman. *Legal Protection for Micro-Biological and Genetic Engineering Inventions*. Reading, Mass.: Addison-Wesley Publishing Co., 1982.

Sanford, Terry. "The University and Technology: New Paths and New Perspectives." In *The Law of Business and Licensing*. Ed. Robert Goldscheider and Tom Arnold. New York: Clark Boardman Co., Ltd., 1982.

Smith, Bruce L. R., and Joseph J. Karlesky. *The State of Academic Science: The Universities in the Nation's Research Effort*. New York: Change Magazine Press, 1977.

Stanford, California. Stanford University Library. W. H. Cowley Papers. "Taxonomy of American Higher Education."

Steward, Irvin. *Organizing Scientific Research for War: The Administrative History of the Office of Scientific Research and Development*. Boston: Little, Brown and Co., 1948.

Storr, Richard J. *The Beginnings of Graduate Education in America*. Chicago: University of Chicago Press, 1953.

Strombler, Milton. *Guide to University–Industry Research Agreements*. Santa Monica, Calif.: Society of Research Administrators, 1983.

Treusch, Paul E., and Norman A. Sugarman. *Tax-Exempt Charitable Organizations*. 2d ed. Philadelphia: American Law Institute/American Bar Association, 1983.

Tribe, Laurence H. *Channeling Technology Through Law*. Chicago: The Bracton Press, Ltd., 1973.

Weeks, K. M., ed. *Legal Deskbook for Administrators of Independent Colleges and Universities*. Notre Dame, Ind.: Center for Constitutional Studies, 1982.

Welzenbach, Lanora F., ed. *College and University Business Administration*. 4th ed. Washington, D.C.: National Association of College and University Business Officers, 1982.

Wescoe, Clarke. "The Use of University Resources." In *The Corporation and the Campus*. Ed. Robert H. Connery. New York: Praeger Publishers, 1970.

Wilson, John T. *Academic Science, Higher Education, and the Federal Government, 1950–1983*. Chicago: University of Chicago Press, 1983.

Wolfle, Dael. *The Home of Science: The Role of the University*. New York: McGraw-Hill, 1972.

York, Herbert F., and G. Allen Greb. "Military Research and Development: A Postwar History." In *Science, Technology and National Policy*. Ed. T. J. Kuehn and A. L. Porter. Ithaca, N.Y.: Cornell University Press, 1981.

Yoxen, Edward. *The Gene Business: Who Should Control Biotechnology?* New York: Harper & Row, 1983.

PERIODICALS AND ARTICLES

Abelson, Philip H. "Academic Science and Industry." *Science* 183 (29 March 1974):1251.

———. "The Federal Government and Innovation." *Science* 201 (11 August 1978):487.

———. "Technology and the U.S. Economy." *Science* 213 (21 August 1981):823.

———. "Prospects for Support of R&D in Fiscal Year 1983." *Science* 216 (2 April 1982):9.

———. "Differing Values in Academia and Industry." *Science* 217 (17 September 1982):1095.

Adams, John F., and John W. Hall. "Legal Liabilities in Higher Education: Their Scope and Management." Section II. *Journal of College and University Law* 3 (1976):244.

Aiken, Ray J. "Legal Liabilities in Higher Education: Their Scope and Management." Section I. *Journal of College and University Law* 3 (1976):181.

"Allies in the Laboratory: University and Industry Scientists Create a Model for Research Agreements." *Washington University Magazine* 53 (Summer 1983):19–22.

American Civil Liberties Union. "Policy Statement of American Civil Liberties Union on University and Contract Research with Emphasis on Growing Ties between Corporations and Academic Institutions." 28 October 1983 (mimeographed).

Anderson, Alison G. "Conflicts of Interest: Efficiency, Fairness and Corporate Structure." *UCLA Law Review* 25 (1978):738–795.

Anderson, David A. "Licensing of College and University Trademarks." *Journal of College and University Law* 8 (1981–82):100.

Arem, Lawrence J. "The Benefits of Research and Development Expenses Increased By New Credit." *Taxation for Lawyers* 10 (January-February 1982):208–213.

Atkinson, Richard C. "Tax Incentives and Research." *Science* 208 (2 May 1980):449.

Auerbach, Lewis E. "Scientists in the New Deal: A Pre-War Episode with Relations Between Science and Government in the U.S." *Minerva* 3 (Winter 1965):457–82.

Bartlett, Joseph W., and James V. Siena. "Research and Development Limited

Partnerships as a Device to Exploit University Owned Technology." *Journal of College and University Law* 10 (1983–84):435–454.

Battenburg, Joseph R. "Forging Links Between Industry and the Academic World." *Journal of the Society of Research Administrators* 12 (Winter 1981):5–11.

Bearn, Alexander A. "The Pharmaceutical Industry and Academe: Partners in Progress." *American Journal of Medicine,* July 1981, pp. 81–88.

Bell, Shela T., and Martin F. Majestic. "Protection and Enforcement of College and University Trademarks." *Journal of College and University Law* 10 (1983–84):65–67.

Beynon, J. H. "Eternal Triangle; Research Universities and Industry." *Chemistry and Industry*, 17 March 1979, pp. 175–82.

Biddle, Wayne. "A Patent on Knowledge: Harvard Goes Public." *Harper's*, July 1981, pp. 22–26.

Black, Robert L. "The Research Credit: R & D in the Post-ERTA Period." *New York University Institute on Federal Taxation* 41 (1983):12–1.

Bok, Derek C. "Universities and National Research Policy." *Science* 193 (10 September 1976):955.

———. "President's Report: Business and the Academy." *Harvard Magazine*, May-June 1981, pp. 23–35.

Bouton, Katherine. "Academic Research and Big Business: A Delicate Balance." *New York Times Magazine*, 11 September 1983, p. 62.

Bremer, Howard W. "Academia—Business Partner." *Les Nouvelles*, September 1979, pp. 155–160.

Broad, William J. "Report Absolves Harvard in Case of Fakery." *Science* 215 (12 February 1982):874–76.

———. "Pajaro Dunes: The Search for Consensus." *Science* 216 (9 April 1982):155.

Caldart, Charles C. "Industry Investment in University Research." *Science, Technology, & Human Values* 8 (Spring 1983):27.

Carey, Susan. "Facing Cuts in Federal Grants, Big Schools Try to Get Research Work from Business." *Wall Street Journal*, 9 February 1982, sec. 2, p. 27(W).

Carey, William D. "Patent Policy Versus Innovation." *Science* 200 (30 June 1978):1439.

Carlot, Louie. "Busch Funding Sparks Biotechnology Research." *Student Life,* 27 September 1983, p. 1.

Cocking, Clive. "Biotechnology: Too Risky for Universities?" *Chronicle of Higher Education,* 29 June 1981, p. 7.

"College and Industry Leaders' Recommendations on Commercial Use of Research." *Chronicle of Higher Education* 24 (7 April 1982):12.

Coolley, Ronald B. "Alternatives to Litigation." *Les Nouvelles*, 18 (September 1983):164.

Cooper, Iver P. "The Patent System and the 'New' Biology." *Rutgers Journal of Computers, Technology and the Law* 8(1) (1980):1–46.

Crawford, Allan R. "Science and Industry." *Science* 213 (4 September 1981):1076–77.

Culliton, Barbara J. "Harvard and Monsanto: The $23–Million Alliance." *Science* 195 (25 February 1977):759–763.

———. "Biomedical Research Enters the Marketplace." *New England Journal of Medicine,* 304 (14 May 1981): 1195–1201.

———. "Pajaro Dunes: The Search for Consensus." *Science* 216 (9 April 1982):155.

———. "The Academic-Industrial Complex." *Science* 216 (28 May 1982):960–62.

———. "The Hoechst Department at Mass. General." *Science* 216 (11 June 1982):1200–1203.

———. "Monsanto Gives Washington U. $23.5 Million." *Science* 216 (18 June 1982):1295–96.

———. "Stanford Doctors Try Consulting, Inc." *Science* 217 (17 September 1982):1122–23.

David, Edward E., Jr. "The University-Academic Connection in Research: Corporate Purposes and Social Responsibilities." *Journal of the Patent Office Society*, April 1982, pp. 209–19.

———. "Supporting Research with a Commercial Mission." *Change* 14 (September 1982):26–29.

Davidson, Duncan M. "Protecting Computer Software: A Comprehensive Analysis." *Jurimetrics Journal* 23 (Summer 1983):337–425.

Davis, Bernard D. "Sounding Board: Profit Sharing Between Professors and the University?" *New England Journal of Medicine*, 14 May 1981, pp. 1232–1235.

Davis, Lance E., and Daniel J. Kevles. "The National Research Fund: A Case Study in the Industrial Support of Academic Science." *Minerva* 12 (1974):207–220.

Dewitt, Donovan J. "Licensing in University Environment." *Les Nouvelles*, December 1976, p. 201.

Dietrich, J. J., and Rajat Sen. "Government–University–Industry Interaction in Research and Development: A Case Study." *Research Management*, September 1981, pp. 23–25.

Dillion, Kristine E., and Karen L. Bane. "Consulting and Conflict of Interest." *Educational Record,* Spring 1980, pp. 52–72.

Doan, Herbert D. "New Arrangements for Industry-Academic Research." *Research Management*, March 1978, pp. 33–35.

Duvall, Jack. "Letter to the Editor." *Chronicle of Higher Education*, 18 May 1983, p. 32.

Ellingsworth, Patrick J. "Investing in Research and Development Limited Partnerships." *National Law Journal* 5 (18 July 1983):38.

———. "The R&D Limited Partnership: When the Research Succeeds." *National Law Journal* 5 (25 July 1983):21.

Finkin, Matthew W. "Regulation by Agreement: The Case of Private Higher Education." *Iowa Law Review* 65 (1980):1120–1200.

Fowler, Donald R. "University–Industry Research Relationships: The Research Agreement." *Journal of College and University Law* 9 (1982–83):515–32.

Fox, Jeffrey. "Can Academia Adapt to Biotechnology's Lure?" *Chemical and Engineering News*, 12 October 1981, pp. 39–44.

Freeman, Alan. "How a Top University Unwittingly Became Stock-Promotion Bait." *Wall Street Journal*, 27 April 1984, p. 1, col. 6.

Fuchs, Ralph F. "Academic Freedom—Its Basic Philosophy, Function, and History." *Law and Contemporary Problems* 28 (1963):431–46.

Fusfeld, Herbert I. "The Bridge Between University and Industry." *Science* 209 (11 July 1980):221.

Gavet, Roy V. "Business and Academia: An Emerging Partnership." *Change* 15 (April 1983):23–28.

Gee, Sherman. "Factors Affecting Innovation Time." *Les Nouvelles*, September 1978, pp. 186–190.

Gibson, G. Thomas. "Research for Sale." *Venture: The Magazine for Entrepreneurs*, March 1984, p. 80.

Gilbert, Lawrence. "Establishing a University Program." *Les Nouvelles*, June 1976, pp. 97–101.

———. "From University to Marketplace." *Les Nouvelles*, December 1977, pp. 285–87.

Goldscheider, Robert, and Michel de Haas. "The Advantages of Arbitration Provisions in License Agreements." *Patent and Trademark Review* 79 (September 1981):347–54.

Gray, Paul E. "M.I.T. Wants Closer Ties With Business." *New York Times*, 18 November 1980, sec. A, p. 1.

Handler, Phillip. "When Science Becomes a Public Venture to Whom is the Scientist Accountable?" *Chronicle of Higher Education*, 3 March 1980, p. 64 (editorial).

Holbert, N. B. "Research: The Ways of Academe and Business." *Business Horizons*, February 1976, pp. 36–38.

Jacobson, Robert L. "Industry Emphasis on Profits Cited as Bar to Business-University Ties." *Chronicle of Higher Education*, 21 July 1982, p. 3.

Johnston, Clement D., and Henry A. Dudley "Conflicts of Interest in Business and Industry." *Federal Bar Journal* 24 (1964):344–50.

Kaplan, Stanley A. "Fiduciary Responsibility in the Management of a Corporation." *Business Lawyer* 31 (1976):886.

Kerr, C. "University and Corporation: Partially Shared Goals and Separate Identities." *California Management Review*, Spring 1974, pp. 20–24.

Kertz, Consuelo L. "Tax Exempt Organizations and Commercially Sponsored Scientific Research." *Journal of College and University Law* 9 (1982–83):69–84.

Kidd, Charles V. "The Implications of Research Funds for Academic Freedom." *Law and Contemporary Problems* 28 (1963):619.

Kiefer, David M. "Forging New and Stronger Links Between University and Industrial Scientists." *Chemical and Engineering News,* 8 December 1980, pp. 38–51.

Kirk, John R., Jr. "Inventions and Patents Resulting From Use of Government Funds: Movement of Government Patent Policy From Diversity Toward Uniformity (Maybe)." *1980 Patent Law Annual* 18 (1980):133.

Kirk, Russell. "Massive Subsidies and Academic Freedom." *Law and Contemporary Problems* 28 (1963):607–12.

Knight, Michael. "Harvard Rules Out Role for Now in Genetic Engineering Company." *New York Times*, 18 November 1980, sec. A, p. 1.

Lachs, Phyllis S. "University Patent Policy." *Journal of College and University Law* 10 (1983–84):263–92.

Lacovara, Philip A. "How Far Can the Federal Camel Slip Under the Academic Tent?" *Journal of College and University Law* 4 (1976):223–240.

Lancaster, Hal. "Efforts to Develop Monoclonal Antibodies Are Moving More

Slowly Than Expected." *Wall Street Journal*, 1 May 1984, sec. 2, p. 37, col. 4.

Lepkowski, Wil. "Academic Values Tested by MIT's New Center." *Chemical and Engineering News*, 15 March 1982, pp. 7–12.

Levin, Doren P. "Westinghouse Move into Robotics Shows Pitfalls of High-Tech Field." *Wall Street Journal*, 14 May 1984, sec. 2, p. 27.

Liszewski, Kathy. "A New Weapon in War on Disease." *St. Louis Post-Dispatch*, 1 June 1982, p. 1–D.

Lohr, Steve. "Campuses Cementing Business Alliances." *New York Times*, 16 November 1980, sec. XII, p. 1, col. 1, cont. p. 25, col. 5.

Lynch, Mitchell C. "Harvard's Plan for Gene-Splicing Company Meets with Scorn from Other Institutions." *Wall Street Journal*, 12 November 1980, sec. 2, p. 37(E).

MacCordy, Edward L. "The Untapped University Resource." *Les Nouvelles* 13 (September 1978):178–85.

McDonald, Kim. "Engineering Schools Urged to Forge Industry Links." *Chronicle of Higher Education*, 14 October 1981, p. 15.

———. "Universities Urged to Bar Secrecy in Pacts with Private Industry." *Chronicle of Higher Education*, 7 April 1982, pp. 1 and 12. *See* accompanying text of Pajaro Dunes Conference statement, pp. 12 and 13, entitled "College and Industry Leaders' Recommendations on Commercial Use of Research."

———. "Washington U., Monsanto Set $25 Million Pact." *Chronicle of Higher Education*, 9 June 1982, p. 9.

———. Congress Asked to Make Researchers Disclose Their Business Ties." *Chronicle of Higher Education*, 23 June 1982, p. 15.

———. "Education, Not Money, Seen as Main Benefit of Research in Industry." *Chronicle of Higher Education*, 30 June 1982, p. 6.

———. "Chemical Company to Spend $6M on Research at MIT." *Chronicle of Higher Education*, 1 September 1982, p. 17.

———. "Science and Industry Discuss Research Partnerships at Universities." *Chronicle of Higher Education*, 5 January 1983, p. 4.

———. "Safeguards Urged in University Agreements with Business." *Chronicle of Higher Education*, 5 July 1984, p. 4.

McLain, L. "Industry and University Must Recognize the Need for Co-operation." *Engineer*, 12 February 1976, p. 37.

Magarrell, Jack. "Academe and Industry Weigh a New Alliance." *Chronicle of Higher Education*, 5 February 1979, pp. 1 and 8.

———. "Corporate Leaders Wary of Federal Aid to Private Universities." *Chronicle of Higher Education*, 2 July 1979, p. 11.

———. "Corporate Chiefs Favor Giving to Higher Education, Survey Finds." *Chronicle of Higher Education* 24 (26 April 1982):7.

Malott, R. H. "Corporate Support of Education: Some Strings Attached." *Harvard Business Review,* July 1978, pp. 133–38.

Mansfield, Edwin. "Tax Policy and Innovation." *Science* 215 (12 March 1982):1365–71.

Marsh, Crispin. "When Agreements Fail." *Les Nouvelles* 18 (September 1983):183.

Meyerhoff, Albert H. "U.C. Research Ties: A Secret Needing a Dose of Sunshine." *Los Angeles Times*, 14 February 1982, sec. 4, p. 5.

Millett, M. L., Jr., and R.B. Holloway. "Research in Industry and on Campus." *Engineering Education*, March 1980, pp. 449–52.

Moore, Nicholas G., Jay F. Ayers, and Frank R. Pope. "How Limited Partnerships Tax-Shelter the R&D of New Products or Technology." *Journal of Taxation* 49 (September 1978):138.

Moss, Thomas H. "New Partnerships a Bargain for Industry, a Boon for Colleges and Universities." *Chronicle of Higher Education*, 6 April 1983, p. 72.

Murray, Thomas J. "Industry's New College Connection." *Dun's Review*, May 1981, pp. 52–54, 59.

———. "The Pitfalls of R&D Partnerships." *Dun's Business Month* 123 (April 1984):127–28.

Nelson, Richard R., and Richard N. Langlois. "Industrial Innovation Policy: Lessons from American History." *Science* 219 (18 February 1983):814–818.

"New Council to Strengthen Research Partnerships." *Chronicle of Higher Education*, 16 May 1984, p. 18.

Noble, David F., and Nancy E. Pfund. "The Plastic Tower: Business Goes Back to College." *Nation*, 20 September 1980, pp. 233, 246–52.

Norman, Colin. "MIT Agonizes Over Links With Research Unit." *Science* 214 (23 October 1981):416–17.

———. "Electronics Firms Plug into the Universities." *Science* 217 (6 August 1982):511–14.

———. "Potential Conflicts of Interest Detailed at UC." *Science* 219 (21 January 1983):268–69.

Note, "Academic Freedom and Federal Regulation of University Hiring." *Harvard Law Review* 92 (1978):881.

Note, "Academic Freedom Privilege: An Excessive Solution to the Problem of Protecting Confidentiality." *University of Cincinnati Law Review* 51 (1982):326–352.

Note, "Developments in the Law—Academic Freedom." *Harvard Law Review* 81 (1967):1049–1159.

Note, "Preventing Unnecessary Intrusions on University Autonomy: A Proposed Academic Freedom Privilege." *California Law Review* 69 (1981):1538–68.

Oaks, Dallin H. "A Private University Looks at Government Regulation." *Journal of College and University Law* 4 (1976):1–12.

———. "Taking Research into the Marketplace." *New England Journal of Medicine*, 9 September 1982, pp. 694–700.

"Pajaro Dunes Conference: Draft Statement." *Journal of College and University Law* 9 (1982–83):534.

Pake, George E. "Some Industrial Perspectives on the University–Industry Relationship." Council of Graduate Schools, *Communicator*, April 1980, pp. 1–2, 8–10. Revised version published in *Physics Today*, January 1981, pp. 44–47.

———. "U.S. Technology in Jeopardy." *Washington University Magazine,* 54 (Spring 1984):18–23.

Patton, Carl V. "Consulting by Faculty Members." *Academe* 66 (1980):181–85.

Patton, Carl V., and James D. Marver. "Paid Consulting by American Academics." *Educational Record*, Spring 1979, pp. 175–84.

Perry, Suzanne. "Professors Can Land Corporate Sponsors for Research—If They Follow the Rules." *Chronicle of Higher Education,* 20 April 1983, p. 21.

Peterson, Donald W. "Industry View of Campus Licensing." *Les Nouvelles,* March 1978, pp. 28–31.

Praeger, Dennis J., and Gilbert S. Omenn. "Research, Innovation, and University Industry Linkages." *Science* 207 (25 January 1980):379–84.

Price, Derek deS. "R&D and Productivity." *Science* 211 (13 March 1981):1116.

"Regulation by Agreement: The Case of Private Education." *Iowa Law Review* 65 (1980):1120–1200.

Reinhold, Robert. "Stanford and Industry Forge a Research Link." *New York Times*, 10 February 1984, sec. A, p. 22.

Roark, Anne C. "Academic Scientists Eye the Industry Connection." *Chronicle of Higher Education,* 30 April 1979, p. 1.

———. "Scientists Back Steps to Improve Return on Research Dollar." *Chronicle of Higher Education*, 23 July 1979, pp. 1 and 10.

Roberts, Edward B., and Donald H. Peters. "Commercial Innovations from University Faculty." *Research Management*, May 1982, pp. 24–30.

"Role of Universities in Industrial Research." *Research Management,* September 1974, pp. 23–37.

Roy, Rustum. "University–Industry Interaction Patterns." *Science* 178 (1 December 1972):955–60.

Rozen, Miriam. "Biotech Firms Offer 'Interim' Products." *Dun's Business Month* 123 (May 1984):103.

Sanger, David E. "U. of California Puts Limits on Private Research Pacts." *New York Times*, 21 August 1983, p. 12.

Sawhill, John. "An Expanded Role for University R&D." *Chemical and Engineering News*, 24 July 1978, p. 5.

Saxon, David S. "University–Industry Teamwork." *Les Nouvelles*, 14 (December 1979):232–35.

Schaffer, Edward. "The University in Service to Technocracy." *Educational Theory* 30 (Winter 1980):50–51.

Scott, Austin W. "The Fiduciary Principle." *California Law Review* 37 (1949):539–55.

Shapiro, Harold T. "Is Taking Sides a Good Idea for Universities?" *Science* 225 (6 July 1984):9.

Shapley, Willis H. "R&D in the Fiscal 1983 Budget." *Science* 217 (16 July 1982):205.

Sheils, Merrill, et al. "Innovation: Has America Lost Its Edge?" *Newsweek*, 4 June 1979, pp. 58–68.

Singer, Dale. "The Academic-Industrial Complex." *St. Louis Post-Dispatch*, 26 September 1982, p. A–13; 27 September 1982, p. 1, col. 1, and p. 6A, col. 1; 28 September 1982, p. Al.

Smith, Arthur A., Jr. "Implications of Uniform Patent Legislation to Colleges and Universities." *Journal of College and University Law* 8 (1981–82):87.

Smith, R. Jeffrey. "Patent Policy Changes Show Concern." *Science* 199 (17 March 1978):1190.

Spikins, Dr. David. "Universities Can Help Industry." *Les Nouvelles*, September 1979, pp. 138–42.

Sproull, R. L. "Industrial-Academic Interface: From the Academic Viewpoint." *Research Management*, May 1969, pp. 219–26.

Stauffer, Thomas M. "Expanding Business—Higher Education Coordination on Research and Development." *SRA Journal*, Summer 1979, pp. 41–46.

Steinbach, Sheldon E. "Establishing an Equitable Patent Policy." *Educational Record*, Spring 1978, pp. 142–47.

Sun, Marjorie. "NIH Ponders Pitfalls of Industrial Support." *Science* 213 (3 July 1981):113–14.

———. "Biotech Firm Biogen is Going Public." *Science* 219 (25 February 1983):938–39.

———. "Scientists Settle Cell Line Dispute." *Science* 220 (22 April 1983):393–94.

Swalin, R. A. "Improving Interaction Between the University and the Technical Community." *Research Management*, May 1976, pp. 25–27.

Teltsch, Kathleen. "Gifts to the Universities Rise 6.2%, Survey Finds." *New York Times*, 6 May 1984, sec. Y, p. 43.

Thomas, Lewis. "Business and Basic Science." *Bulletin of the New York Academy of Medicine* 57(6) (July-August 1981):493–502.

Thordahl, Monte C. "The Importance of Managing Innovation." *Les Nouvelles*, March 1977, pp. 48–50.

"Trading Jobs in Academia and Industry Helps High-Technology Firms." *Wall Street Journal*, 26 June 1984, p. 1.

Udell, Gerald G.; Ken Baker, and Robert M. Colton. "The Innovation Center Program." *Research Management*, July 1979, pp. 32–38.

Udell, Gerald G., and Richard D. Johnson. "Campus Attitudes Change." *Les Nouvelles*, September 1978, pp. 175–77.

"Universities Look to Industry for Support." *Chemical and Engineering News*, 12 April 1982, pp. 16–18.

Useem, Michael. "Business Segments and Corporate Relations with U.S. Universities." *Social Problems* 29 (December 1981):129–41.

Vernon, Raymond. "Gone Are the Cash Cows of Yesteryear." *Harvard Business Review*, November- December 1980, pp. 150–55.

Wade, Nicholas. "Cloning Gold Rush Turns Basic Biology into Big Business." *Science* 208 (16 May 1980):688–92.

———. "Three New Entrants in Gene Splicing Derby." *Science* 208 (16 May 1980):690–692.

———. "Harvard Marches Up Hill and Down Again." *Science* 210 (5 December 1980):1104.

Waldholz, Michael. "R&D Partnerships Proliferate,Offer Investors a Tax Shelter Plus a Shot at High-Tech Profits." *Wall Street Journal*, 26 September 1983, sec. 2, p. 60.

Walker, David. "British Universities Step Up Research Parks to Foster Ties with Technology Companies." *Chronicle of Higher Education*, 16 March 1983, p. 19.

Walsh, J. "Unfaculty a Growing Factor in Research." *Science* 204 (20 April 1979):286.

Walsh, John. "What Can Government Do for Innovation?" *Science* 205 (27 July 1979):378–80.

Watkins, Beverly T. "Outside Income on University Time: A Conflict for Professors?" *Chronicle of Higher Education*, 2 June 1982, p. 14.

Weber, David. "A New Industry Springs to Life." *Venture*, May 1981, pp. 88–93.

"Weighing University Research Proposals." *Chemical Week*, 3 February 1982, pp. 55–56.

Weston, Michael C. " 'Outside' Activities of Faculty Members." *Journal of College and University Law* 7 (1980–81):68–77.

White, P. C. "Better Industry–University–Government Cooperation—Why and How?" *Research Management*, January 1973, pp. 10–15.

Willenbrock, F. Karl. "United States and Technological Preeminence." *Science* 213 (18 September 1981):1319.

Wilson, John W., Jr., Richard P. Dobb, and William T. Gerl. "Consider Various Important Factors When Contracting for University R & D." *EDN*, 31 March 1983, p. 298.

Winer, Jonathan M. "Law Profs: Their Hidden Income." *National Law Journal*, 17 September 1979, pp. 1 and 10.

Woerpel, Marvin D. "Research to Realization Via Campus." *Les Nouvelles*, December 1976, pp. 197–98.

Wolff, Michael. "The President's Initiatives for Industrial Innovation." *Research Management*, January 1980, pp. 7–12.

———. "The Why, When, and How of Directed Basic Research." *Research Management,* May 1981, pp. 29–31.

Zettlemoyer, A. C. "Industry/Academic Interface." *Chemical and Engineering News*, 12 March 1979, p. 36.

SYMPOSIA AND REPORTS

Business-Higher Education Forum. *Corporate and Campus Cooperation: An Action Agenda*. Washington, D.C.: Business-Higher Education Forum, May 1984.

Fowler, Donald R. "Conflicts of Interest." Paper presented at NACUA Mid-Winter CLE Workshop, Washington, D.C., 11–12 March 1983.

Stam, T. L. "Fundamentals of Patent Policy: Patent Policy for Educational Institutions." In *Technology Transfer: University Opportunities and Responsibilities. A Report on the Proceedings of a National Conference on the Management of University Technology Resources*. Cleveland, Ohio: Case Western Reserve University, 1974.

Stanford University. *Policy on Tangible Research Property*, 16 March 1982.

———. *Policy on Consulting by Members of the Academic Council: Principles and General Standards*, 2 December 1982.

Stanford University and the University of California. *A Cooperative University/Industry Program in Biotechnology*. Fact Sheet issued in December 1981.

Twentieth Century Fund, Task Force on the Commercialization of Scientific Research. *The Science Business*. New York: Priority Press, 1984.

University of California Systemwide Administration, Office of the President. *Interim Guidelines on University–Industry Relations: Issues and Recommended Responses,* 3 November 1982.

University Council on Biotechnology. "Patent Task Group Report." Chapel Hill, N.C., March 1982 (mimeographed).

UNPUBLISHED MATERIALS

Fowler, Donald R. "A Study of the Need for and the Impediments to Improved and Novel University–Industry Research Relationships." Ph.D. dissertation, Claremont Graduate School, 1983.

Fried, Dr. Josef, Louis Bloch Professor of Chemistry at the University of Chicago. Telephone interview, 26 July 1983.

Kipnis, David, Washington University School of Medicine, St. Louis, Missouri. Interview, 16 April 1984.

MacCordy, Edward L., Washington University, St. Louis, Missouri. Interview, 29 February 1984.

MacCordy, Edward L., and Harry S. Leahey, Letter to Dr. Joseph Davie, 18 August 1983.

Orf, Harry W., Massachusetts General Hospital, Boston, Mass. Interview, 10 April 1984.

Reams, Bernard D., Jr. "Research Interactions Between Industry and Higher Education: An Examination of the Major Legal Issues Involved in Four Representative Contracts." Ph.D. dissertation, Saint Louis University, 1984.

Rosenzweig, Robert M., Letter to William Danforth, 10 January 1984.

Ruscio, Kenneth P. "University Industry Relations in Biotechnology: A Study of the Public Policy Issues." Ph.D. dissertation, Syracuse University, 1983.

Thompson, Larry R., Letter, 8 July 1983.

Tolbert, Tom L., Monsanto Company, St. Louis, Missouri. Interview, 25 June 1984.

CASES

Agulnick v. American Hospital Supply Corp., 507 F. Supp. 135 (D.C. Mass. 1981).

Armstrong Paint & Varnish Works v. Nu-Enamel Corp., 305 U.S. 315 (1938).

Barenblatt v. United States, 360 U.S. 109 (1959).

Barnes v. Lone Star Steel Co., 642 F.2d 993 (5th Cir. 1981).

Bob Jones University v. United States, 103 S. Ct. 2017 (1983).

Brown v. Wichita State University, 217 Kan. 279, 540 P.2d 66 (1975).

City of Coral Gables v. Weksler, 164 So.2d 260 (Fla. 1964).

Commercial Union Ins. Co. v. Ford Motor Co., 640 F.2d 210 (9th Cir. 1981).

Community Communications v. Boulder, 455 U.S. 40 (1982).

Cooper v. Ross, 472 F. Supp. 802 (E.D. Ark. 1979).

Cornell University v. Messing Bakeries, 135 N.Y.S.2d 101, *modified*, 285 A.D. 490, 138 N.Y.S.2d 280 (1955), aff'd, 309 N.Y. 722, 128 N.E.2d 421, *re-argument denied,* 309 N.Y. 800, 130 N.E.2d 601 (1955).

Diamond v. Chakrabarty, 447 U.S. 303 (1980).

Dow Chemical v. Allen, 672 F.2d 1262 (7th Cir. 1982).

E. I. Du Pont de Nemours Powder Co. v. Maslund, 244 U.S. 100 (1917).

E. L. White, Inc. v. City of Huntington Beach, 146 Cal. Rptr. 614, 579 P.2d 503 (1978).

Energy Resources Corp. v. Porter, 14 Mass. App. 296, 438 N.E.2d 391 (1982).

First Equity Corp. of Florida v. Utah State University, 544 P.2d 887 (Utah 1975).

Franchised Stores of New York, Inc. v. Thompson-Hudson Co., et al., 300 F. 509 (6th Cir. 1924).
Goldberg v. Medtronic, Inc., 686 F.2d 1219 (7th Cir. 1982).
Gross v. University of Tennessee, 448 F. Supp. 245 (W.D. Tenn. 1978).
Hanover Star Milling Co. v. Metcalf, 240 U.S. 403 (1916).
Helmrich v. Eli Lilly & Co., 455 N.Y.S. 460 (1982).
Hill v. Okay Const. Co., 252 N.W.2d 107 (1977).
In re Heard, 6 B.R. 876 (Bankr. Ky. 1980).
Jones & Laughlin Steel v. Johns-Manville Sales, 453 F. Supp. 527 (W.D. Pa. 1978).
Keyishian v. Bd. of Regents of the State of New York, 385 U.S. 589 (1967).
Krotkoff v. Goucher College, 585 F.2d 675 (4th Cir. 1978).
Lafayette v. Louisiana Power & Light Co., 435 U.S. 389 (1978).
Link-Belt Co. v. Star Iron & Steel Co., 65 Cal. App. 3d 24, 135 Cal. Rptr. 134 (1976).
McClane v. Sun Oil Co., 634 F.2d 855 (5th Cir. 1981).
Meinhard v. Salmon, 249 N.Y. 458, 164 N.E. 545 (1928).
Mink v. University of Chicago, 460 F. Supp. 713 (N.D. Ill. 1978).
Moore v. Regents of the University of California, Case No. C513755 (September 12, 1984).
New Amsterdam Casualty Co. v. Waller, 233 N.C. 536, 64 S.E.2d 826 (1951).
President & Trustees of Colby College v. Colby College–N.H., 508 F.2d 804 (1st Cir. 1975).
Randall v. Bd. of Higher Education in the City of New York, 78 A.D.2d 516, 432 N.Y.S.2d 164 (1980).
Regents of the University of California v. Bakke, 438 U.S. 265 (1978).
St. Luke's Hospital of Kansas City v. United States, 494 F. Supp. 85 (W.D. Mo. 1980).
Smith v. Seaboard Coast Line F. Co., 639 F.2d 1235 (5th Cir. 1981).
Speck v. North Carolina Dairy Foundation, Inc., 307 S.E.2d 785 (N.C. App. 1983), rev'd. 319 S.E.2d 139 (N.C. 1984).
Stern v. Lucy Webb Hayes Nat. Train. School for Deaconesses and Missionaries, 381 F. Supp. 1003 (D.D.C. 1974).
Sweezy v. New Hampshire, 354 U.S. 234 (1957).
United Drug Co. v. T. Rectanus Co., 248 U.S. 90 (1918).
University of Notre Dame Du Lac v. J. C. Gourmet Food Imports Co., 703 F.2d 1372, 217 U.S.P.Q. 505 (Fed. Cir. 1983).
University of Notre Dame Du Lac v. Twentieth Century–Fox Film, 15 N.Y.2d 940, 207 N.E.2d 508 (1965).
University of Pittsburgh v. Champion Products, 686 F.2d 1040 (3d Cir. 1982), *cert. denied*, ______U.S.______ 103 S. Ct. 571 (1982), Civ. A. No. 81–2167 (W.D. Pa., 23 June 1983).
Vitamin Technologists, Inc. v. Wisconsin Alumni Research Foundation, 146 F.2d 941 (9th Cir. 1945).
Wetherill v. University of Chicago, 518 F. Supp. 1387 (N.D. Ill. 1981); 548 F. Supp. 66 (1982); 565 F. Supp. 1553 (1983); 570 F. Supp. 1124 (1983). Annot. 2 A.L.R.4th 1091 (1980).

STATUTES AND ADMINISTRATIVE RULES AND REGULATIONS

Act donating Public Lands to the Several States and Territories which may provide Colleges for the benefit of Agriculture and Mechanical Arts. Ch. CXXX, 12 Stat. 503 (1862).

Act of 29 March 1984, 1984 Ga. Laws 1196.

Army Appropriations Act. Ch. 418, sec. 2, 39 Stat. 649 (1916).

26 C.F.R. Part 1 (proposed amendments), 48 Fed. Reg. 2790 (1983).

Cal. Administrative Code. Title 2, sec. 18705.

Cal. Corp. Code. Sec. 5111.

Cal. Corp. Code. Sec. 5232.

Cal. Corp. Code. Sec. 5240.

Cal. Corp. Code. Sec. 5260.

Federal Grant and Cooperative Agreement Act of 1977. 92 Stat. 3 (1978).

47 Fed. Reg. 7561, 7582 (1982).

Florida Administrative Code. Ch. 6C1, sec. 7.392.

Lanham Act. 15 U.S.C. Sec. 1051 et seq. (1976).

Local Government Antitrust Act of 1984. 98 Stat. 2750 (1984).

National Cooperative Research Act of 1984. 98 Stat. 1815 (1984).

National Defense Education Act. 72 Stat. 1580 (1958).

Order by the Council of National Defense Establishing the National Research Committee, 27 June 1940, 5 Fed. Reg. 2446 (1940).

Patent and Trademark Laws Amendment of 1980. 94 Stat. 3015, 3019–3029 (1980).

Rev. Rul. 76–296, 1976–2 C.B. 142.

Stevenson-Wydler Technology Innovation Act of 1980. 94 Stat. 2311 (1980).

Treas. Reg. Sec. 1.501 (c)(3)–1(d) (5) (1) (1976).

See also Rev. Rul. 296, 1976–2 Cum. Bull. 142–143.

Treas Reg. Sec. 1.174–2 (a) (1981); Rev. Rul. 65–1, 1965–1 Cum. Bull. 226.

Uniform Deceptive Trade Practices Act. 7A U.L.A. 35, 39 (1968).

Uniform Patent Legislation. 35 U.S.C. Sec. 220 et seq. (as amended 1980).

U.S. *Constitution.* Art. I, sec. 8.

———. President. Exec. Order No. 2859, 11 May 1918.

———. President. Message to Congress on Science and Technology, 8 Weekly Comp. Pres. Doc. 587 (16 March 1972).

Wis. Stat. Sec. 270.58 (1971) (current supp.).

LEGISLATIVE MATERIALS AND GOVERNMENT DOCUMENTS

Advisory Committee on Industrial Innovation. *Direct Federal Support of Research and Development: Draft Report.* Washington, D.C.: Industrial Innovation Coordinating Committee, 21 December 1978 (NTIS # PB–290 407).

———. *Domestic Policy Review, Final Report.* Washington, D.C.: U.S. Department of Commerce, 1979.

Baruch, Bernard M. *American Industry in the War: A Report of the War Industries Board*. Washington, D.C.: U.S. Government Printing Office, 1921.

Beidenbach, J. M. "Currently-Used Mechanisms and Delivery Modes for Nonacademic Personnel." *Continuing Education in Science and Engineering*. National Science Foundation, SE78-5, 1977, pp. 91–123.

Burger, R. M. *Analysis of the National Science Foundation's Innovation Centers Experiment*. Springfield, Va.: National Technical Information Service, 1977.

———. *Analysis of the National Science Foundation's University–Industry Cooperative Research Centers Experiment*. Springfield, Va.: National Technical Information Service, 1979.

Bush, Vannevar. *Science—The Endless Frontier: A Report to the President on a Program for Postwar Scientific Research*. July 1945.Washington, D.C.: U.S. Government Printing Office, 1945.

Cantwell, Katherine M. "University–Industry Research Relationships at the Stanford Synchrotron Radiation Laboratory." A report submitted to the National Science Board, July 1980.

Circular No. A–124, Office of Management and Budget. Patents—Small Business Firms and Nonprofit Organizations, 47 Fed. Reg. 7559–7566 (1982).

Commissioner of Patents and Trademarks: Annual Report, Fiscal Year '81. Washington, D.C.: U.S. Government Printing Office, 1982.

Department of Justice. *Antitrust Guide Concerning Research Joint Ventures*. Washington, D.C.: U.S. Government Printing Office, November 1980.

Economic Recovery Tax Act of 1981. 95 Stat. 172–355 (1981) 26 U.S.C.A. secs. 1–7701(a)(34)(A) passim (1982).

Gilpin, Robert. *Technology, Economic Growth, and International Competitiveness*. A report prepared for the Subcommittee on Economic Growth of the Joint Economic Committee, Congress of the United States, 9 July 1975.Washington, D.C.: U.S. Government Printing Office, 1975.

Grad, Marcia L. "An Analysis of the Cooperative R&D Concept." Vol. II, Report R77–02, National Science Foundation Contract No. C–860.Prepared for Industrial Program, Applied Science and Research Applications Directorate, National Science Foundation, October 1977.

———. "Evaluation Status and Planning II, Cooperative Research and Development Experiment." Report R77–03, Industrial Program, Applied Science and Research Applications Directorate, National Science Foundation, November 1977.

MacCordy, Edward L. "Prospects for Government/University/Industry Research Cooperation." Paper presented before the Division of Science Resource Studies, National Science Foundation, Washington, D.C., 22 September 1980.

Mogee, Mary Ellen. "The Relationship of Federal Support of Basic Research in Universities to Industrial Innovation and Productivity." Washington, D.C.: Library of Congress, Congressional Research Service, 24 August 1979.

———. "The Relationship of Federal Support of Basic Research in Universities to Industrial Innovation and Productivity." In U.S. Congress, Joint Economic Committee, Special Study on Economic Change. Volume 3.*Research & Innovation: Developing a Dynamic Nation*. Washington, D.C.: U.S. Government Printing Office, 29 December 1980, pp. 257–79.

Myers, Sumner, and Donald G. Marquis. *Successful Industrial Innovations: A Study of Factors Underlying Innovation in Selected Firms.* Washington, D.C.: National Science Foundation, 1969.

Myers, Sumner, and Eldon E. Sweezy. *Federal Incentives for Innovation: Why Innovations Falter and Fail; A Study of 200 Cases.* Report R–75–04.Washington, D.C.: National Science Foundation, 1976.

Nason, Howard K., and Joseph A. Steger. *Support of Basic Research by Industry.* Washington, D.C.: National Science Foundation, 1978.(Prepared for NSF/STIA/SRS under Grant NSF C–76–21517.)

National Academy of Sciences. *Annual Report for 1918.* Washington, D.C.: U.S. Government Printing Office, 1919.

———. *Forum on Technological Innovation and the U.S. Economy.* Washington, D.C.: 14 November 1978.

National Resources Committee. "Relation of the Federal Government to Research." *Research—A National Resource.* Pt. 1. Washington, D.C.: U.S. Government Printing Office, 1938.

———. "Industrial Research." *Research—A National Resource.* Pt. 2.Washington, D.C.: U.S. Government Printing Office, 1941.

Office of Technology Assessment. "Applications of R&D in the Civil Sector." Office of Technology Assessment, June 1978.

———. *Government Involvement in the Innovation Process.* Washington, D.C.: U.S. Government Printing Office, 1978.

Place, Geoffrey. "The Government Role in the Development and Commercialization of Technology." In *Science and Technology Annual Report to the Congress.* National Science Foundation, June 1980, Volume II of the ASTR Commissioned Paper Series.

Shapero, Albert. *Report on University-Industry Interactions: Recurring Expectations, Unwarranted Assumptions and Feasible Policies.* Ohio State University for National Science Foundation, 31 July 1979.

Socolar, M. J. Comptroller General of the United States. Letter addressed to the Honorable Albert Gore, Jr., 16 October 1981, ref. B–204687.

Steelman, John R. *Science and Public Policy: A Program for the Nation.* Vol. I. Washington, D.C.: U.S. Government Printing Office, 1947.

Teich, Albert H., W. Henry Lambright, and Vivien B. Shelanski. "The Role of National Laboratories in University–Industry Relationships." A report to the National Science Board, September 1981.

Teichl, Judith L. "Inventory of University–Industry Research Support Agreements in Biomedical Science and Technology." A report prepared for the National Institute of Health, 1982.

Thackray, Arnold. "University–Industry Connections and Chemical Research: An Historical Perspective." A report submitted to the National Science Board, 1982.

Throdal, M. C. "The University-Monsanto Research Interface." In *Proceedings, Engineering and Science Research for Industrial Development.* Ed. Fred Landis. Washington, D.C.: National Science Foundation, October 1977, pp. 390–96.

Tornatzky, Louis G., et al. *University–Industry Cooperative Research Centers: A*

Practice Manual. Washington, D.C.: National Science Foundation, May 1982.

U.S. Congress. House. Committee on Banking, Finance and Urban Affairs. Subcommittee on Economic Stabilization. *Forging an Industrial Competitiveness Strategy*. 98th Cong., 1st sess., 1983. Committee Print 98–8.

———. Congress. House. Committee on the Budget. *Higher Education and Innovation in the U.S. Economy and President's Fiscal Year 1984 Budget: Perspective from the States. Hearings before Task Force on Energy and Technology and Task Force on Education and Employment*. 98th Cong., 1st sess., 1 March 1983.

———. Congress. House. Committee on Foreign Affairs. *Overview of U.S. International Competitiveness. Hearings before a Subcommittee of the House Committee on Foreign Affairs on Aerospace Industry, Machine Tools Industry, and Electronics Industry*. 97th Cong., 2d sess., 1983.

———. Congress. House. Committee on Government Operations. *Research and Development (Part 1—Background Testimony; Part 2—Military Research Representatives). Hearings before a Subcommittee of the House Committee on Government Operations*. 85th Cong., 2d sess., 1958.

———. Congress. House. Committee on the Judiciary. *Industrial Innovation and Patent and Copyright Law Amendments. Hearing before a Subcommittee on Courts, Civil Liberties, and the Administration of Justice*. 96th Cong., 2d sess., 3 April–9 June 1980.

———. Congress. House. Committee on Science and Technology. *Government and Innovation: University–Industry Relations. Hearing before a Subcommittee on Science, Research, and Technology*. 96th Cong., 1st sess., July 31, August 1, 2, 1979.

———. Congress. House. Committee on Science and Technology. *The Government-University Accountability Relationships in the Field of Scientific Research. Hearing before the Subcommittee on Science, Research, and Technology*. 96th Cong., 2d sess., March 24, 1980.

———. Congress. House. Committee on Science and Technology. *National Engineering and Science Policy. Hearings before a Subcommittee of the House Committee on Science and Technology*. 97th Cong., 1st sess., December 14, 1981.

———. Congress. House. Committee on Science and Technology. *Commercialization of Academic Biomedical Research. Hearings before a Subcommittee on Science, Research and Technology*. 97th Cong., 1st sess., June 8, 9, 1981.

———. Congress. House. Committee on Science and Technology. *University–Industry Cooperation in Biotechnology. Hearing before the Subcommittee on Investigations and Oversights*. 97th Cong., 2d sess., 16, 17 June 1982.

———. Congress. House. Committee on Science and Technology. *Robotics. Hearings before the Subcommittee on Investigations and Oversights*. 97th Cong., 2d sess., 2, 23 June 1982.

———. Congress. House. Committee on Ways and Means. *General Revenue Revision. Hearing before the Committee on Ways and Means on Topics Pertaining to the General Revenue Revision of the Internal Revenue Code. Part 2*. 85th Cong., 2d sess., 1958.

———. Congress. Joint Hearing before the Senate Committee on Commerce,

Science, and Transportation and the House Committee on Science and Technology. *Federal Patent Policy*. 97th Cong., 1st sess., 30 September 1981.

———. Congress. Office of Technology Assessment. *Commercial Biotechnology: An International Analysis*. Washington, D.C.: U.S. Government Printing Office, OTA-BA-218, 1984.

———. Congress. Senate. Committee on Governmental Affairs. *Research and Development in the United States: The Role of the Public and Private Sectors. Hearing before a Subcommittee on Energy, Nuclear Proliferation, and Government Processes*. 97th Cong., 2d sess., 9 March 1982.

———. Congress. Senate. Committee on Military Affairs. Subcommittee on War Mobilization. *The Government's Wartime Research and Development 1940–44*. Subcommittee Report No. 5. Washington, D.C.: U.S. Government Printing Office, 1945.

———. Congress. Senate. Committee on Military Affairs. Subcommittee on War Mobilization. *Report on Science Legislation: National Science Foundation*. Subcommittee Report No. 8. Washington, D.C.: U.S. Government Printing Office, 1946.

———. Congress. Senate. *National Technology Innovation Act of 1979*. S. Rept. 96–781. 96th Cong., 2d sess., 15 May 1980.

———. Department of Commerce. "Cooperative Generic Technology Program Procedures." *Federal Register*. Vol. 45, no. 159, 14 August 1980, pp. 54028–54031.

———. Department of Commerce. General Counsel for Economic Affairs. "Information and Steps Necessary to Form Research and Development Limited Partnerships." 30 November 1982 (mimeographed).

———. Department of Commerce. Office of Cooperative Technology. "Cooperative Technology Program: Key to Industrial Innovation." Revised Draft, 24 January 1980.

———. Department of Commerce. *Information and Steps Necessary to Form Research and Development Limited Partnerships*. November 1982.

———. Department of Commerce. Office of Productivity, Technology and Innovation. Office of Cooperative Generic Technology. *Cooperative R&D Programs to Stimulate Industrial Innovation in Selected Countries*. Washington, D.C.: various dates in 1979 and 1980.

———. Department of Justice. *Antitrust Guide Concerning Research Joint Ventures*. November 1980. Washington, D.C.: U.S. Government Printing Office, 1980.

———. Food Administration. *Annual Report for the Year 1918*. Washington, D.C.: U.S. Government Printing Office, 1919.

———. General Accounting Office. *The Federal Role in Fostering University–Industry Cooperation*. Washington, D.C.: General Accounting Office, 1983.

———. National Science Board. *Science Indicators—1980*. Washington, D.C.: U.S. Government Printing Office, 1981.

———. National Science Board. *University–Industry Research Relationships: Myths, Realities, and Potentials, Fourteenth Annual Report*. Washington, D.C.: U.S. Government Printing Office, 1982.

———. National Science Foundation. *An Analysis of the National Science Foun-*

dation's Innovation Centers Experiment. Washington, D.C.: U.S. Government Printing Office, 1979.

———. National Science Foundation. *Development of University–Industry Cooperative Research Centers: Historical Profiles*. Washington, D.C.: Innovation Processes Research Section Division of Industrial Science and Technological Innovation, National Science Foundation, May 1982.

———. National Science Foundation. *Federal Support to Universities, Colleges, and Selected Non-Profit Institutions: A Report to the President and Congress*. Washington, D.C.: U.S. Government Printing Office, 1981.

———. National Science Foundation. *National Patterns of Science and Technology Resources 1982*. Washington, D.C.: U.S. Government Printing Office, 1982.

———. National Science Foundation. *Proceedings of a Conference on Academic and Industrial Basic Research*. Princeton University, November 1960.(NSF 61–39).

———. National Science Foundation. *Research in Industry: Roles of the Government and the National Science Foundation*. Washington, D.C.: National Science Foundation, December 1976.

———. National Science Foundation. *Thirty-First Annual Report for Fiscal Year 1981*. Washington, D.C.: U.S. Government Printing Office, 1982.

Index

About the Author

BERNARD D. REAMS, Jr., is Professor of Law at Washington University in St. Louis, where he also serves as Director of the Freund Law Library. He is the author of many books, including *Federal Price and Wage Control Programs 1917–1979; Congress and the Courts: A Legislative History 1787–1977* with C. R. Haworth; and *Human Experimentation: Federal Laws, Legislative Histories, Regulations and Related Documents* with C. J. Gray. Reams is a member of the Kansas and Federal bars.